SPINNING WHEELS
AND ACCESSORIES

David A. Pennington
and Michael B. Taylor

4880 Lower Valley Road, Atglen, PA 19310 USA

Dedication

Those of us who collect know what a disease collecting can be, as basements, attics, small outbuildings and even barns fill up. Over the past century several collectors have made significant contributions to chronicling the history of the spinning wheel. John Horner, Samuel Dale Stevens, Marion Channing, Joan Cummer and others have been patient amassers of great collections, which have become invaluable aids to those who came along later. We dedicate this book to them and all who have shared our passion for spinning wheels.

Library of Congress Cataloging-in-Publication Data

Pennington, David A.
   Spinning wheels and accessories / by David A. Pennington and Michael B. Taylor
     p. cm.
1. Spinning-wheel. I. Taylor, Michael B. II. Title.
TS1484 . P46 2004
677'02822-dc22
          2003023077

Copyright © 2004 by David A. Pennington and Michael B. Taylor

All rights reserved. No part of this work may be reproduced or used in any form or by any means—graphic, electronic, or mechanical, including photocopying or information storage and retrieval systems—without written permission from the publisher.
The scanning, uploading and distribution of this book or any part thereof via the Internet or via any other means without the permission of the publisher is illegal and punishable by law. Please purchase only authorized editions and do not participate in or encourage the electronic piracy of copyrighted materials.
"Schiffer," "Schiffer Publishing Ltd. & Design," and the "Design of pen and ink well" are registered trademarks of Schiffer Publishing Ltd.

Designed by Ellen J. (Sue) Taltoan
Type set in Windsor BT/Aldine 721 BT

ISBN: 0-7643-1973-6
Printed in China

Published by Schiffer Publishing Ltd.
4880 Lower Valley Road
Atglen, PA 19310
Phone: (610) 593-1777; Fax: (610) 593-2002
E-mail: Info@schifferbooks.com
For the largest selection of fine reference books on this and related subjects, please visit our web site catalog at
www.schifferbooks.com
We are always looking for people to write books on new and related subjects. If you have an idea for a book, please contact us at the above address.

This book may be purchased from the publisher.
Include $3.95 for shipping. Please try your bookstore first.
You may write for a free catalog.

In Europe, Schiffer books are distributed by
Bushwood Books
6 Marksbury Ave. Kew Gardens
Surrey TW9 4JF England
Phone: 44 (0)20 8392-8585; Fax: 44 (0)20 8392-9876
E-mail: info @Bushwoodbooks.co.uk
Free postage in the UK. Europe: air mail at cost.
Please try your bookstore first.

Contents

Acknowledgments 4
Introduction 5
1. The Basics 7
2. Wool Wheels 14
3. American Saxonies 30
4. "Unusual" Flax Wheels 40
5. American-Irish Castle Wheels 53
6. Double Treadle, Double Wheel Flax Wheels 60
7. Double-Flyer Wheels 79
8. Accelerating Wheel-heads: "The Pleasant Spinner," Amos Minor,
    and the "Minor's Head" 90
9. Patent Wheels 97
10. Famous Makers 114
11. Eastern Pennsylvania Saxony Makers 123
12. Shaker Wheels 131
13. Fancy European Wheels 140
14. Spinning Accessories 159
15. Tips On Collecting 188
Appendix. Spinning Wheel Makers—Marks and Information 190
Glossary 217
Bibliography 219
Index 223

Acknowledgments

Our first book, *A Pictorial Guide to American Spinning Wheels*, now out of print, was very successful, but the requirements of raising families, pursuing careers, and collecting spinning wheels left us without the time to do more than write articles about the field. We kept promising ourselves, and others, to do a second book. Administrators at Marietta College were encouraging to Mike, but it never happened. Then in the summer of 2001, we got serious because Cassie Cowan, a young, 20th century American historian, playfully insisted that Mike "stop sitting around taking up space." Mike then went to Dave and said that it was time to get moving, and for some reason it actually happened.

James Munsie has been helping us out since the 1970s, when he shot the first book's photographs. He coached Dave on all aspects of photography for this book and allowed us to show wheels from his collection. William Leinbach, the noted collector/dealer from Myerstown, Pennsylvania, has been feeding, sheltering, and sharing information with Mike since 1986. Jeanne Asplundh, another long-time collector/dealer, has a magnificent collection and has been wonderful about sharing information, hosting and housing Mike, and arranging for her collection to be photographed for this book. Florence Feldman-Wood, the publisher of *The Spinning Wheel Sleuth*, has provided on-going email encouragement for about ten years. Her journal has been a source of much information and a venue for our articles, which kept our hands and heads in the game. The Shakers at Sabbathday Lake, Maine, were gracious enough to publish the first book. Sister R. Mildred Barker and Brother Theodore Johnson were wonderful, and we are grateful to them and the current members of the community.

Beth Pennington, Dave's wife, has been patiently feeding, refereeing literary disputes, dusting and carrying wheels during photographing sessions, and keeping our spirits up as we have plodded along. Our children, Lynna and Karl Pennington and Will and Hallie Taylor, have endured our long-time hobby/obsession/sickness with good humor and patience. They have even encouraged us to "buy another one."

Obviously, over the 30 years that we have been working on "the second book," many people have helped us, and we thank them, one and all.

Spinning Wheel Prices

Assigning prices to any antique is a difficult task, even when the object is right in front of you for examination. To fully convey the condition and attributes of the piece in a brief caption is impossible, and hence taints the entire pricing process. We have tried to give the current market prices of the complete or nearly complete examples shown. We have discovered, over the last 30 years, that wheel prices also vary geographically, due to the local availability of a particular type or style. Condition, completeness, style, execution, maker's mark, and relative rarity all play a part in establishing a fair price. The one thing that can safely be said about price guides is that the minute they reach print they are out of date. Please note that these prices were current in 2003. If they make you pleased with your purchase or possession, we are delighted. If you are angry and wish to complain about the price we placed on your wheel, you now know why we didn't include our home phone numbers here.

Introduction

For many years spinning wheels have been an icon for the antiques industry. They are symbols of an earlier age when people were more self-reliant. Single women were "spinsters" who supported themselves through their spinning efforts. Women were referred to as the "distaff set" after that part of the spinning wheel, the distaff, which held the flax fibers prior to their being spun into linen thread. At one time distaffs were so common in the antiques trade they sold for $0.25 and a dealer made a fence of spinning wheels over a quarter of a mile long next to his shop. Now some prime examples have sold for thousands of dollars at public auction. By the 1920s some collectors were beginning to take notice of spinning wheels, not sure whether to focus on them as furniture or tools. In 1917 Wallace Nutting included them in his *Furniture Treasury*. In 1929 an early issue of *The Magazine Antiques* had an article on unusual forms of spinning wheels. In the late 1950s the heirs and friends of Samuel D. Stevens, textile industrialist and pioneer collector, opened a museum to continue his pioneering work on American textile tools. That museum evolved into The American Textile History Museum in Lowell, Massachusetts, with a collection of hundreds of wheels thanks to the accession of the collections of Anne Pixley and Joan Cummer, which augmented that of Stevens. In 2000 the Home Textile Tool Museum opened in Orwell, Pennsylvania (mailing address: RR1, Box 141A, Rome, PA 18837). Books about spinning wheels by Marion Channing, Patricia Baines, and Joan Cummer, as well as numerous books about spinning and taking care of spinning wheels, have appeared since the 1960s. In 1993 *The Spinning Wheel Sleuth* (now a quarterly), devoted to unusual forms of spinning wheels, came on the scene. In 1975 we published a book, *A Pictorial Guide to American Spinning Wheels*, which went through three printings and 15,000 copies over the next 23 years. Now we are returning to this fascinating tool, which occasionally rose to the level of fine furniture. Information about spinning wheels in the American colonies is quite fragmentary and very speculative. We also have included some European examples of the fanciest sort, as they bring exceedingly high prices and are dramatically different, as well as closely related spinning tools, such as reels and swifts.

Writing about the spinning wheel is difficult because it is a tool made by turners and other furniture craftsmen. The text must explain how the wheels function as tools and how some of the forms they take are technologically interesting, while still acknowledging that occasionally some examples are considered furniture. It is not uncommon to find fine turnings, bone inlay, and figured woods used on American wheels. Wheels were commonly stained and even painted to resemble mahogany.

Our primary goal is to use words and many pictures to help readers understand their antique American wheels. We hope you read the text as we have sweated bullets over it, but we have tried also to include as much information as possible with the pictures and captions.

There have been a host of revival spinning wheels, some extremely beautiful and clever, over the last 100 years; but they are not our subject, nor are the wheels imported by the container load from Europe and other countries. We will show some wheels that were made in immigrant communities in the United States in the late 19th century, which are very similar to recent imports. In such cases only wood analysis can finally determine from where a wheel comes.

We also plan to tell the story of some prominent late 18th and early 19th century spinning wheel makers. They literally left their marks on their wheels. They were highly skilled craftsmen, often also producing chairs and other items of furniture. They are listed in census records as turners, joiners, cabinetmakers, and even wheelwrights. The spinning wheel was not something commonly made by "Pa" out in the barn, no matter how talented a jack-of-all trades he might have been. Given the shop records of some makers, "Pa" wasn't even allowed to repair them. Of course, this complicates the organization of the book because some makers or groups of makers (the Shakers for example) produced enough wheels in enough variety to warrant separate treatment.

After introducing basic wheel forms and terminology, the book commences with chapters that introduce the common wool wheels and flax wheels in their typical forms, giving examples from the major regions, with an emphasis on the "old North." Next, we present the odd wheels, sometimes known in patent records and sometimes not. Then we discuss some of the great makers who have distinguished themselves by leaving us enough interesting wheels and documentation to warrant special consideration. Finally, we consider some of the associated textile tools, which are both beautiful and interesting, including some European wheels of great artistry and clever technology. Appendix A is a list of over 1000 wheel makers with their marks, when and where they worked, and what kinds of wheels they made.

When we were writing our first book in the early 1970s, container loads of western European wheels and truck loads of French-Canadian wheels were flooding the U.S. market. People buying wheels needed some way of identifying early American wheels, as well as unusual wheels. Now, with on-line auctions and importers bringing wheels from eastern Europe in quantity, prospective buyers again need some help. We offer this book to old friends who found our first work helpful and to new friends who would like some information about their wheel or guidance when looking for an old one. A few people are actually spinning wheel collectors, but that curse is relatively rare. One writer about antiques, George Grotz, warned his readers that spinning wheel collecting was a disease with 99% of the populace immune, but the 1% had a real problem. If he had written .000001%, he would have been more accurate, but the warning still holds. We salute the incurable!

Measurements Are Absent

Without getting into a great debate (probably endless as well as inconclusive) we will simply relate that five serious collectors, each having collected wheels for over 30 years, have measured countless wheels and found no useful recurring measurements for understanding who, where, or when spinning wheels were made. So, individual wheel measurements are *not* a part of this book; we have included a 3" block scale in the photos to give approximate sizes. We also give the circumference of the drive wheel (DW), and on one class of wheels powered by a rope and pulley system, i.e. "Pleasant Spinners," we give the circumference of the accelerating wheel (W).

The common wool and flax wheels dating from about 1775 to 1900 and used in America comprise the majority of the examples here, but we also encounter spindle wheels that are over eleven feet long as well as some less than two feet in their largest dimension. We will even see spindle wheels without a drive wheel.

Wheel showing 3 inch black scale. See page 91 for details on this wheel.

Chapter 1

The Basics

Spinning wheel terminology has varied over the years. The Glossary, at the back of this book, includes terms we use and others that might be handy. Various proposals for terminology reform have been floated, but none seem to have stuck. Not being reformers, we will use terms widely employed on the American antique scene.

Fig. 1-1 and Fig. 1-2 show the two common types of American wheels with labeled parts. In most cases, spinning wheels, or "wheels," have a *drive wheel*, the turning of which supplies power through a *drive cord*, or band, to turn either a *spindle* or a *bobbin and flyer* unit, which imparts a twist to the fiber being spun. The drive wheels may be turned by hand or with a foot-powered *treadle* and *crank* arrangement. The crank, on the drive wheel axle, and the foot treadle are connected by the *footman*, or pitman. American wheels usually have a *table* with *legs* to bring the spinning apparatus up to a convenient height for the spinner. The drive wheel is suspended from one or two *wheel supports (posts)* on an *axle*. We call the unit for keeping the proper tension, on the drive cord, the *tension device*.

The two common types of spinning wheels are the *wool wheel* and the *flax wheel*. In general, we refer to all spindle wheels as wool wheels, and all bobbin and flyer wheels as flax wheels. Of course, either type can be used to spin any fiber, including silk, wool, flax, cotton, hemp or even modern synthetics. Flax wheels have been called "small wheels," "linen wheels," or just plain "spinning wheels" by their makers. Wool wheels have been referred to as "great wheels," "spindle wheels," "long wheels," "large wheels," "muckle wheels," and "walking wheels."

Wool Wheels

Wool wheels (Fig. 1-1) were technological improvements to the hand-turned spindle and seem first to have appeared in Asia by 1200 CE. An American wool wheel has a large drive wheel (about 44 inches in diameter) turned by hand with the spinner standing next to the drive wheel. The purpose of the large wheel is to allow the spinner to reach and turn the wheel while walking away from the spindle during the spinning process. On the wool wheel the spinner must periodically stop spinning and in a separate action wind the spun thread onto the spindle. Many wool wheels have a hole in the top of the spindle post into which a *wheel head* is inserted. The wheel head holds the spindle and in some cases an accelerating wheel to increase the wool wheel's efficiency. Rims on American wool wheels are almost all *hoop* rims. They are made of one or two splits of wood, one to four inches wide, somewhat narrow and thicker if there is a groove, and somewhat wider and thinner if not. The splits are steamed or well soaked, so that they will bend suitably to make a wheel 44 to 48 inches in diameter. In parts of Europe and Asia there were rimless wheels. A cord or ribbon ran diagonally from spoke to spoke to form a kind of rim around which a drive band passed. In only one instance have we seen a rimless wool wheel which we think is American in origin.

WOOL WHEEL

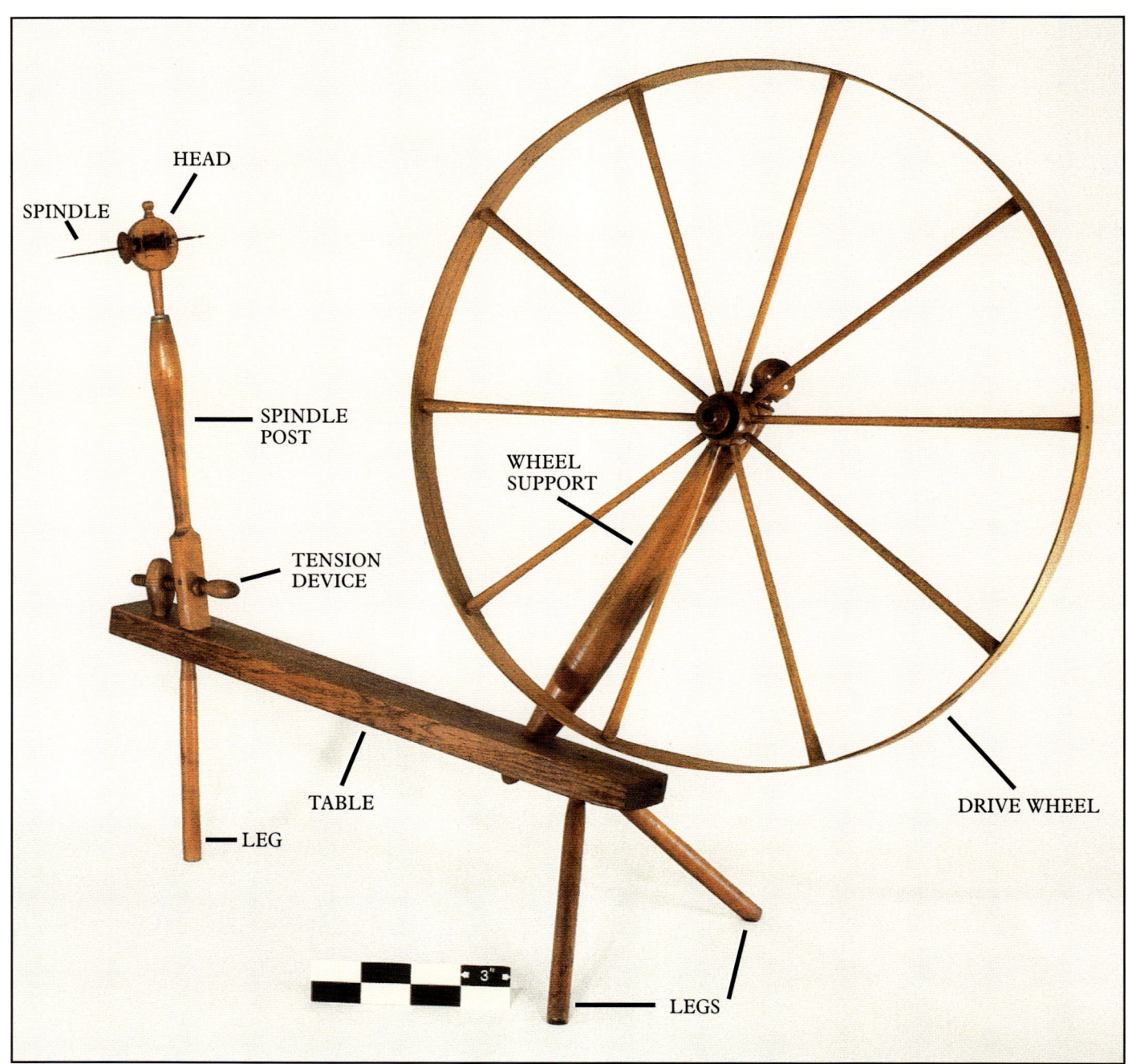

Fig. 1-1. American wool wheel with labeled parts. DW=44". See the Chapter 2 on wool wheels for the variety of prices on American wool wheels. *Courtesy of James Munsie.*

FLAX WHEEL

Fig. 1-2. American saxony with labeled parts. DW=21". See Chapter 3 on saxony wheels for the variety of prices on American saxonies. *Courtesy of James Munsie.*

Fig. 1-3. Five bobbin and flyer units to give some sense of the variety. From left to right: Daniel Danner, Manheim, Pennsylvania Irish castle wheel; Thomas Cushman Alfred, Maine Shaker community saxony wheel; unique assembly from unusual New York state double wheel, double treadle wheel; Holland, Michigan, Dutch-American saxony wheel; and "I. Homsher", Montgomery County, Pennsylvania saxony wheel. *Courtesy of Jeanne Asplundh.* $75-100.

Fig. 1-4. Two bobbin and flyer units disassembled. The one on the left is typical except that it has the hooks on the top right of the flyer arms. The one on the right is unique and shows an example of American ingenuity as well as for comparison. *Courtesy of Jeanne Asplundh.*

Flax Wheels

People in northern climes put legs under the table to get themselves up off the cold floor. By the late 15th century the bobbin and flyer wheel was developed and the typical three-legged flax wheel with a treadle and crank arrangement, commonly referred to as a "*saxony*," emerged (Fig. 1-2). [The story of European wheels is carefully covered in Patricia Baines' book, *Spinning Wheels, Spinners, and Spinning*.] On a flax wheel a *bobbin and flyer* unit (see Fig. 1-3) is suspended from *leathers* between two *maidens*. These maidens sit in a *mother-of-all* made up of several pieces, which can be moved away from the drive wheel by means of a tension screw at the end of the table to tighten the drive band. Flax wheels commonly have a foot treadle, especially on American versions. Wool wheels are turned by hand, normally while the spinner is standing. The bobbin and flyer unit has four pieces, three of which are separable (see Fig. 1-4). The typical American *flyer* is made up of a metal shaft with wooden arms affixed to the shaft. The *bobbin* is a wooden spool with a pulley on the end. It rotates freely on the metal shaft. The last part is a wooden *whorl*, which is threaded and removable. In the spinner's end of the shaft is a forged hole or *orifice*. The orifice protrudes through the front leather. Just behind the orifice a U-shaped pair of wooden *arms* are affixed to the shaft. Along the top edge of the arms are metal *hooks*, also called *teeth*, nailed into the arms. On most American flax wheels, the hooks are on the left arm as one looks down from above the wheel. The far end of the metal shaft is normally threaded and the wooden flyer whorl with an inset metal nut is thereby fastened to the shaft. (ALERT! The threading on the flyer whorl thread is "left-handed," so when removing the flyer whorl, be careful to turn it in the correct direction, i.e. "backwards." Occasionally, the shaft is tapered so that the whorl is fastened with a "jam fit.") The bobbin whorl is a different diameter (almost always smaller) than the flyer whorl. It appears that there are two drive cords, but it is really one that goes around the drive wheel twice and drives both the flyer whorl and the bobbin whorl. The difference in the whorl sizes causes the spun fiber to be drawn onto the bobbin as the spinner works. There are three types of bobbin and flyer unit, but the one described here is by far the most common on American wheels. Bobbin and flyer units both spin the thread and wind it onto the bobbin. They do both by having either the bobbin or the flyer go around faster. If the bobbin goes around faster, it is called a "bobbin lead." If the flyer goes faster, it is a "flyer lead." The doubled drive cord found on most American and many European wheels is one kind of bobbin lead.

On many wheels from Alpine regions, only the bobbin is driven and a "brake" is applied to the flyer (Fig. 1-5). That is another type of bobbin lead. The third main type is a "flyer lead," which employs a brake on the bobbin. This type is much rarer. One of group saxonies using this system are called Picardy wheels because of their prevalence in the 18th century in the Picardy region in France (Fig. 1-6). Very old French-Canadian saxonies have this arrangement too. A number of small upright wheels in Germany and Czechoslovakia from the late 19th century also employ this system.[1]

Normally a flax wheel will have a distaff assembly consisting of three parts. The top piece is actually the *distaff*. The lower two sections are appropriately called the *distaff holder*. Often on flax wheels there is a *peg* or pegs to keep the axle of the drive wheel steady in the slotted ends of the wheel support posts. The drive wheels of most saxonies are quite substantial and made from four solid pieces of wood fitted together and fastened with pegs, or splines. On a few early flax wheels we find hoop rims, sometimes called "bentwood" rims, similar to the sort commonly found on wool wheels.

Fig. 1-5. Swiss style flax wheel with bobbin lead arrangement. It has a metal flyer.
Heavily imported since the 1960's. DW=18½".
Courtesy of Jeanne Asplundh. $350-400.

Fig. 1-6. Close up of a Picardy bobbin and flyer unit with a flyer lead. The bobbin and flyer are outside of the two maidens instead of inside. The wheel is shown in Fig. 3-2.

Makers' Identification

A *maker's mark* is a name or set of initials stamped, burned, or punched with a small pick, into an available flat surface on a wheel, usually the table. (In the Appendix a list of makers and their marks appears, along with information about the types of wheels they made and where and when they were working.) Occasionally an applied decal was used, or sometimes a paper label was affixed. If a wheel is marked, we show that information in the caption and in some cases we have illustrated them. Examples occur throughout the book.[2] Initials picked into the table may be those of the recipient rather than the maker; however, there are examples of makers who picked their names or initials. By the early 18th century Scottish and Irish law required makers of textile tools to put their initials on items they made. While they may have carried the custom to this country, the purpose seems to have changed. Spinning wheel makers often put their wheels out for sale with general merchants. Putting their names on their wheels had some promotional value and gave the customer some reassurance. Identical wheels with and without marks are known. Initially, we thought these were copies by other less reputable makers, but now we suspect that when a wheel was sold directly by the maker, they may not have bothered with the mark. That seems to be the case with the Shakers and with Alpheus Webster. Others, who dated and numbered their wheels such as Samuel Humes and Daniel Danner, were probably concerned about promotion primarily. Obviously, a wheel by a well-known maker is more valuable, whether marked or not, than one about which nothing is known, but beware of those who tell you that "this wheel is just like" those of a well-known maker if they are trying to sell the wheel unless they have the visual evidence to support the claim.

Understanding the Mechanism

Anyone collecting wheels should learn to spin, so they can understand the condition of an antique wheel; for, unfortunately, pieces are commonly missing. Also, antique dealers have been known to "marry" drive wheels to wheel bases or replace bobbin and flyer units, so it is important to see how well these "fit." Unless a bushing has fallen out or there is obvious wear from countless hours of use, the drive wheels should fit properly. Bobbin and flyer units should fit between the leathers such that the whorls line up with the drive wheel and the unit turns freely. Distaffs and distaff holders should have the same turnings as the maidens and/or the wheel posts and should fit.

Saxonies

American saxonies are somewhat larger than their Low Dutch/Low Irish ancestors. Some European saxonies are quite small, but even the slightly larger Low Dutch wheels are noticeably smaller than their American counterparts (Fig. 3-1). We have put small "scales" in our photos, so that size differences between wheels can be better appreciated. We have also indicated the size of the principal drive wheel. Drive wheels on American wool wheels are usually between 44 and 48 inches in diameter, and drive wheels on American saxonies are closer to 20 inches. It is no wonder that people acquiring a spinning wheel as a decorator item want a flax wheel. One very common form of flax wheel, which was not widely made in America, is what we call a "parlor wheel," that is, a small, vertical spinning wheel with the drive wheel below the bobbin and flyer unit. In Fig. 1-7 we show a fancy example, which is a cousin to the wheels in Chapter 13.

Fig. 1-7. This form of vertical flax wheel, which we call a parlor wheel, is commonly found in Europe in all quality levels. This one has a heart-shaped base and sulfur inlays on the rear of the drive wheel. DW=15 ½". *Courtesy of Jeanne Asplundh.* $1800-2000.

---

[1] In Patricia Baines, *Spinning Wheels, Spinners and Spinning*, (New York: Charles Scribner's Sons, 1977): 112-115 there is an excellent discussion of Picardy wheels, including an illustration showing how the system works.. G.B. Thomson, comp. *Spinning Wheels, (The John Horner Collection)*, (Ulster Museum, 1964): 10-15 has a pretty good section on different bobbin and flyer systems, but the illustration and discussion of the Picardy system is incorrect.

[2] Recording and cataloging initials has been a problem because sometimes the period is included in the stamp and sometimes not, for example F.W., not FW. Sometimes, it is D·M, not D.M. or DM. When dealing with initials which were picked with a sharp tool rather than a stamp, the problems increase. In listing them in a data base it made more sense to treat them as if they all had periods and to assume that they were the first and last initials of one name. We know too that the same initials were used by different people, and we assume that if the nature of the mark is different, for example raised rather than incised, the maker is different. Check the Appendix for a list of 1000+ makers and marks.

Chapter 2
Wool Wheels

This chapter provides a quick tour of regional characteristics of wool wheels. The "gallery of wool wheels" at the end of this section visually conveys, better than any text, the similarities and differences among the regions. In wool wheels there are many ways to adjust the tension on the drive cord, and these tension devices are often good clues to the region in which the wheels were made. In addition, legs and wheel-spoke turnings can be helpful with wool wheels as well as flax wheels for making educated guesses about origins. Conclusions on regional characteristics derive from observations of wheels in museums, antiques stores, private collections, reading what others have published, and looking at signed wheels about which there is good documentation.

Tensioning Devices

Tension on the drive cord is increased or decreased by moving the spindle post toward or away from the drive wheel. Very infrequently, the spindle post is fixed. The most common form of tensioning device on wool wheels is shown in Fig. 2-1. It is a threaded rod running though the spindle post, which is free to move. The threaded rod engages a receiver fixed into the wheel table behind the spindle post. Turning the threaded rod one way pushes the spindle post away from drive wheel and increases tension on the drive cord. Turning the rod the other way allows the spindle post to move back toward the drive wheel and relaxes the tension. This type of tension system, a variant of the system commonly found on flax wheels, is easier to use and allows more precise adjustment of the tension than most other types, and hence it is found on wool wheels from New England, the Mid-Atlantic, and the Midwest. On such wheels, turnings on spokes and legs may be helpful clues as to origin.

Fig. 2-1. A typical 19th century New England wool wheel with very plain turnings. It has the most common form of tension screw system. Shown with removable "bat's head" spindle assembly. DW= 47 1/2". *Courtesy of David Pennington.* $225-300.

The wool wheel in Fig. 2-1 is typical of those found throughout New England. The spokes, legs, spindle post, and wheel post, are quite plain. Wool wheels with a tension screw and this plain are likely to be from New England. Wool wheels often have a hole in the top of the spindle post for a removable head. There are several types of removable spindle holders, which we have shown in Fig. 2-2, 2-3, and 2-4. Those from New England usually had either a "bat's head" or an accelerating head.

Fig. 2-2. Five removable "bat's head" spindle assemblies. Most frequently found in New England, but also found in Pennsylvania. *Courtesy of David Pennington.* $50-75 each.

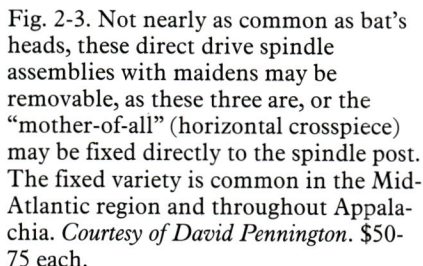

Fig. 2-3. Not nearly as common as bat's heads, these direct drive spindle assemblies with maidens may be removable, as these three are, or the "mother-of-all" (horizontal crosspiece) may be fixed directly to the spindle post. The fixed variety is common in the Mid-Atlantic region and throughout Appalachia. *Courtesy of David Pennington.* $50-75 each.

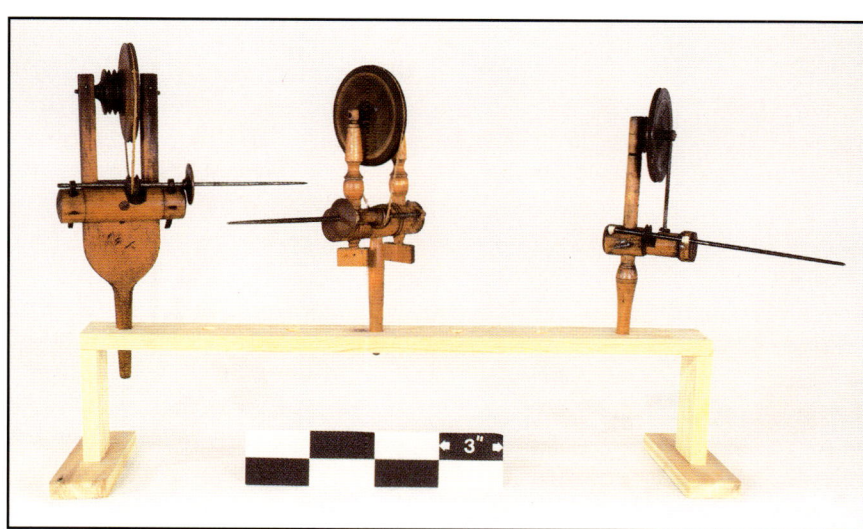

Fig. 2-4. Accelerating heads like these were made in the thousands throughout the 19th century. The "Minor's head" type in the center is quite common. The other two are local variations. *Courtesy of David Pennington.* $75-125

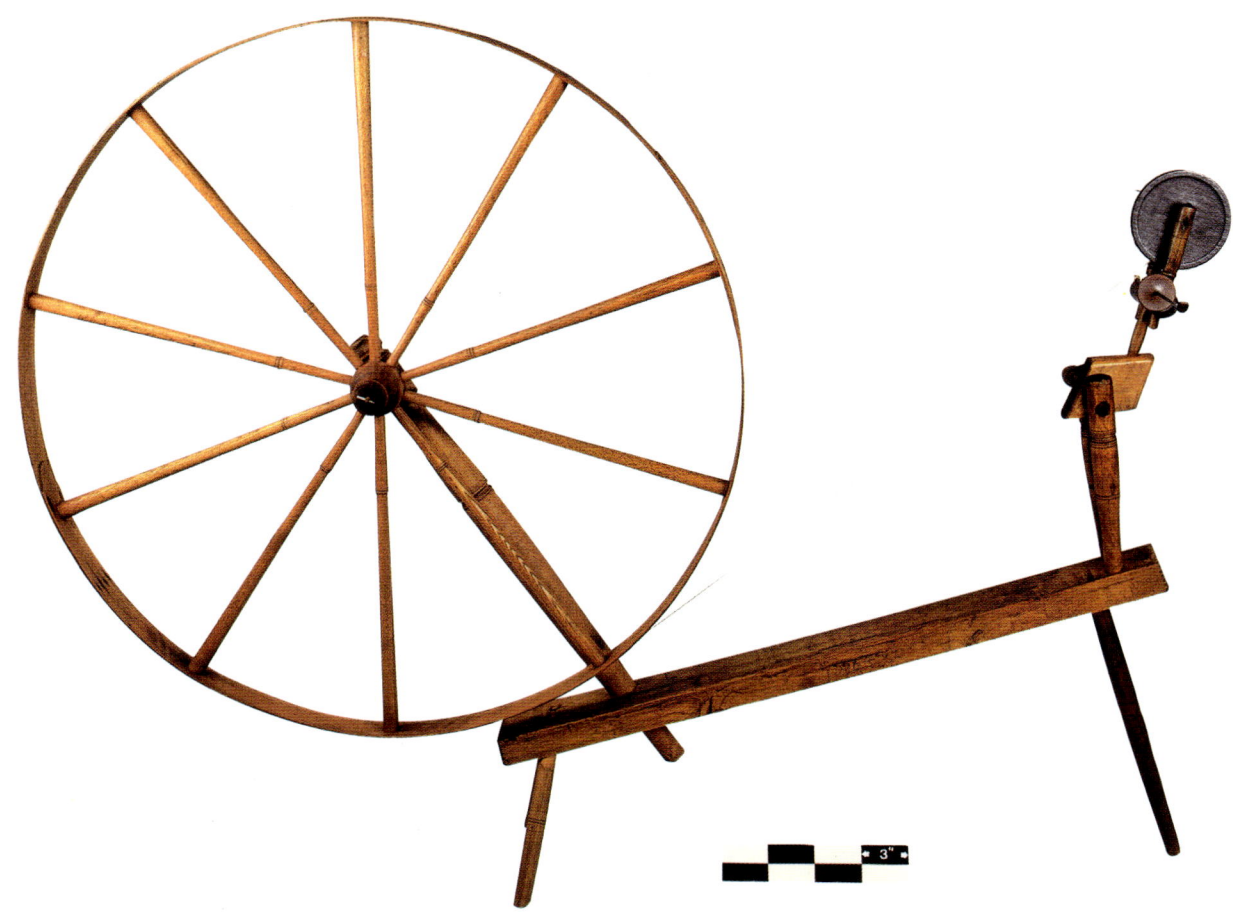

Fig. 2-5. Having a rotating "barrel" (block in this case), this "left-handed" wheel is reversed in its orientation to the spinner, but is otherwise typical of a wool wheel from the Owego, New York area. "Left-handed" Minor's heads also exist suggesting that some preferred turning the wheel with left hand rather than the right. DW=45". *Courtesy of Jeanne Asplundh.* $500-600 in the left-handed version, $350-450 otherwise.

Many New York and Connecticut wool wheel makers favored the rotating barrel tension device seen in Fig. 2-5. The rotating barrel has a hole drilled in it to accept the removable head. The rotating barrel is held in its proper position by a threaded nut putting pressure on the maidens. The foot on the leg on this one is very much like those seen on New York flax wheels.

Settlement patterns help explain the dominant wheel styles in an area. Occasionally, of course, anomalies arise. In Washington County, Ohio, rotating barrel tension devices are not uncommon on wool wheels, as one of the prolific makers of them was a gunsmith named John Vincent. His wife was from upstate New York, perhaps explaining his choice.[1] The Shakers at Pleasant Hill, Kentucky, used a rotating barrel tension device on their octagonal posted wool wheels.

Another very common form of tension system on wool wheels is the sliding table. The wheels in Fig. 2-6 and 2-7 are examples. Both also have fixed direct drive spindle holders. The wheel post is inset into a second table, which rides along above the main table. There is a wooden nut, which can be tightened to hold the sliding table at a point where the tension on the drive band is acceptable. Normally, the tightening nut is above the table, but on the very elaborate wheel shown in Fig. 2-6 the nut is a beautiful carved wooden "wing" nut secured below the main table. The wheel in Fig. 2-6 is quite refined and has little horns and a scooped area to hold rolags, bundles of prepared, unspun fiber, ready for the spinners use. While more than one of these beautiful wheels has been seen, their place of origin can only be guessed at, but the tension device, the long, graceful foot, the relatively heavy wheel post and the fixed, direct drive spindle holders suggest urban eastern Pennsylvania. The turnings on the legs in Fig. 2-7 are common on wool and flax wheels from Connecticut, Pennsylvania and Ohio. It was found in Holmes County, Ohio, and others featuring the octagonal post and sliding table tension system were also seen there.

Fig. 2-6. Extraordinary turnings and details on this wool wheel with sliding table tension system suggest an eastern Pennsylvania urban origin. DW=43". *Courtesy of David Pennington.* $800-900.

Fig. 2-7. An especially nice example of the standard sliding table tension wool wheel found in Pennsylvania and regions settled by its settlers moving westward and southward. It is from Holmes County, Ohio. DW=48". *Courtesy of Michael Taylor*. $400-500.

The "D. Kunkel" wool wheel is quite typical of eastern Pennsylvania wool wheels and was probably the work of Daniel Kunkel of Berks County, Pennsylvania. Its wheel rim is about 4" wide.

Fig. 2-8. Signed "D. Kunkel" and in perfect condition. This eastern Pennsylvania (Berks County) wool wheel with wide rim (4") and massive turnings is almost as good as it gets in wool wheels. Fig. 2-26 is as good as it gets. DW=44". *Courtesy of Michael Taylor*. $800-900.

The wheel in Figs. 2-9 and 2-10 is probably from the Shenandoah Valley, Virginia. The wheel rim is only about 2" wide, and there is a groove in it to hold the drive cord on. The tension system is shown disassembled.

Fig. 2-9. A Virginia wool wheel which uses a wedge under the table rather than a nut and screw to hold the sliding table in place to maintain the proper tension of the drive cord. *Courtesy of Jeanne Asplundh*. DW=46 ½". $600-700.

Fig. 2-10. A view of the sliding table tension system of the Virginia wool wheel disassembled. *Courtesy of Jeanne Asplundh*.

Regional Characteristics

The steep slant of the table of the Virginia wheel in Fig. 2-11 is found on many Appalachian wool wheels, as is the narrow, grooved wheel rim.. Most New England wool wheels have wide, flat rims. In other regions there seems to be more of a mix between the two types of wheel rims. In Tennessee there are very distinctive wool wheels which we have dubbed "Granny Greer" wheels after a photo taken in the 1930s of a "Granny Greer" at work on such a wheel.[2] In Fig. 2-12 we see a fine example and recognize that the tension system is simply a modification of that found on a flax wheel.

Fig. 2-11. Another version of the Virginia wool wheel, this example has the narrow rim with a groove, but looks very different from the preceding example because of its extremely slanted table. This combination of extreme table slant and narrow, grooved rim is found throughout the Appalachian highlands. *Courtesy of David Pennington.* DW=42 ½". $500-600.

Fig. 2-12. Extremely rare and highly sought after, this "Granny Greer" type wool wheel has magnificent turnings and an unusual tension system based on the traditional saxony flax wheel. DW=45". *Courtesy of David Pennington.* $900-1000.

Another very distinctive type of wool wheel, which is probably from Michigan or Ontario, is shown in Fig. 2-13 with its unusual turned table and single post, rotating barrel tension device. A third unusual wool wheel type Fig. 2-14 has reversed the positioning of the legs by placing the two legs at the high end of the table. Such wheels have been found from Rhode Island northward to Canada. The wool wheel in Fig. 2-14 also has other interesting features worth noting. The legs are not turned; they are chamfered and roughly octagonal. The drive wheel axle is wooden rather than metal. While some wheels of this type have a normal screw-type tension device, this one has no apparent tension system at all. The necessary adjustment might be made by simply turning the spindle post in its hole in the table. If the drive cord is fairly tight and not prone to stretching, such a system will work. Perhaps this example is Canadian, but New England is also a reasonable guess.

Fig. 2-13. Probably from Ontario, but possibly Michigan, this wool wheel is distinctive with its round, turned table and its rotating barrel attached directly to the spindle post. DW=50". (noticeably larger than the standard 44-48 inches range encountered). *Courtesy of David Pennington.* $300-400.

Fig. 2-14. Probably from Ontario or Quebec, this wheel has reversed legs, wooden wheel axle, and no tension device. Tension adjustment can be achieved by rotating the spindle post slightly which increases the drag on the drive cord, but it is a very unrefined system. One of the chamfered octagonal legs has been replaced with a shaped tree branch. Wooden axles were once thought to be evidence of great age, but iron was quite scarce at times in rural areas even in the early 19th century. DW=41 ½". *Courtesy of Jeanne Asplundh.* $200-250.

Three tension devices found on Shaker wheels may well have been borrowed from surrounding local wheel culture, so we show the tension devices here as well as in the Shaker chapter. The first in Fig. 2-15 is seen on wheels marked "D.M." from the New Lebanon, NY Shaker community. Here the tension screw pushes the bottom of the spindle post. Wool wheels marked "T.C." and "SR AL" from the Alfred, Maine Shaker community have the tension device shown in Fig. 2-16. In this type a wooden screw presses against a wooden plate to hold the wheel post in its proper place. This type is not nearly as good a system as that shown in Fig. 2-1, and the Shakers at Alfred, Maine changed in 1823 after merchants handling their wares requested that they use the latter system like their Shaker brethren at Canterbury, New Hampshire.[3] The third system (Fig. 2-17) is an under-the-table threaded screw system several of examples of which are in the study collection at the Sabbathday Lake, Maine Shaker community.

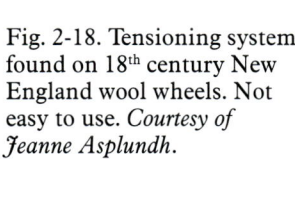

Fig. 2-17. Tensioning system found on Shaker wool wheels from Sabbathday Lake or Poland Hill, Maine. *Courtesy of David Pennington.*

Another tension system occasionally seen in New England on wool wheels is shown in Fig. 2-18. This system is found on wool wheels which often appear quite old, and we suspect the system was not commonly used in the 19th century. It does not allow for incremental tension adjustments.

In Kentucky we have seen a number of examples of another tension system using a threaded rod. The spindle post is fixed into the table, but the bottom half is thinned out so that it is quite flexible. We call it the "green-stick" tensioning system. A tall post behind it acts as the receiver for the threaded rod which when turned causes the post to move away from or toward the wheel post (Fig. 2-19).

Fig. 2-15. Tensioning system on wool wheel marked "D.M." for David Meacham, trustee at New Lebanon, New York Shaker community. *Courtesy of David Pennington.*

Fig. 2-18. Tensioning system found on 18th century New England wool wheels. Not easy to use. *Courtesy of Jeanne Asplundh.*

Fig. 2-16. Tensioning system found on Alfred, Maine Shaker wool wheels marked "T.C." and some "SR AL". Also found on some non-Shaker wheels with wooden axles. *Courtesy of David Pennington.*

Fig. 2-19. This arrangement we call a "green-stick tensioning system". It is commonly found on wheels in Appalachian areas.

Children's Wool Wheels

Small wool wheels are sometimes referred to as children's wool wheels, which makes some sense. Full-sized wool wheels have a spindle height which could be a little difficult to reach for a child. The likely reason for the creation of such wheels would be as a special present, and the one shown in Fig. 2-20 would certainly qualify as such. It even has a bead on the wheel rim, a decorative conceit not even found on nice adult-sized wool wheels. Notice the size difference in the comparison shot in Fig. 2-21.

The child's wheel shown in Fig. 2-22 has an unusual handle-like tension nut. A full-sized wool wheel of this sort has been reported. It makes sense that someone would make both full-sized and child-sized ones of the same style. Most small wheels are quill or bobbin winders sold by the unscrupulous or unknowledgeable as children's wheels for big bucks. Quill or bobbin winders usually have a small peg on one of the wheel's spokes to facilitate rapid turning. In Fig. 2-23 we see an example of such a bobbin winder.

Ephrata, Pennsylvania, the home of the Ephrata Cloisters, is the site of an early 18th century communitarian society of Germans. At the museum there are two small rimless wool wheels which are uncannily similar to one in Patricia Baines' book from that time period in Switzerland.[4] These wheels have cords running diagonally from stave to stave. The drive cord rests on this cord. These are fascinating examples of how immigrants initially made what they knew from "the old country."

Fig. 2-20. A superb example of a child's wool wheel from Pennsylvania. DW=29". *Courtesy of Michael Taylor.* $800-900.

Fig. 2-21. Only by placing a full-sized wool wheel and a child's size together can the difference in size be appreciated.

---

[1] Allen H. Eaton, *Handicrafts of the Southern Highlands*. (New York: Russell Sage Foundation, 1937): 89. Granny Greer was 100-101 years old when the photo was taken on her porch by the legendary chronicler of Appalachia, Doris Ullman. Her full name was Nancy Osborne Greer. In Sadye T. Wilson, and Doris F. Kennedy, *Of Coverlets: the legacies, the weavers*, (Nashville, Tennessee: Tunstede, 1983): 46 it says that Nancy, a native of Johnson County, Tennessee, went to weave in people's homes in Virginia and North Carolina for up to a month at a time. Her wheel is not the one commonly found in Tennessee. Wilson and Kennedy's magnificent book has old pictures of Tennessee weavers, a few of which are shown with their wheels. The dominant type seems to be the direct drive, sliding table tensioning type. Some of the tables are so steeply slanted as to require a wheel support post, Wilson and Kennedy, 143. There is a section on spinning and weaving in Eliot Wigginton's, *Foxfire 2*, (Garden City, Anchor Press/ Double day, 1973): 172-255, which includes pictures of several wool wheels from the Rabun Gap, Georgia area. They have the steep slant to the table and the narrow rim with a single groove which are characteristics of many Appalachian wool wheels.

[2] William Reynolds, "Some More Thoughts on the Vincent Spinning Wheel," *The Association of Ohio Long Rifle Collectors*, Vol. IV, Nancy G. No. 1 (February, 1982): 14-15.

[3] In addition to looking at the chapter on Shaker wheels, the reader should refer to Michael B. Taylor, "Spinning Wheel Study Expands," *The Shaker Messenger*, (Spring 1986): 8-11, 23-27. The distinctive marks found on Shaker wool wheels are pictured, and the story of their production is well covered there.

[4] Patricia Baines, *Spinning Wheels, Spinners and Spinning*, (New York: Charles Scribner's Sons, 1977): 51.

Fig. 2-22. Another wonderful example of a Pennsylvania child's wool wheel. DW=32 ½". *Courtesy of Jeanne Asplundh.* $800-900.

Fig. 2-23. A very nice bobbin winder showing the peg on one spoke to allow for rapid turning when filling them. Such winders are occasionally misrepresented as children's wheels. DW=26". *Courtesy of Jeanne Asplundh.* $250-300.

Gallery of Wool Wheels

The wool wheels in this Gallery are not inferior to the ones illustrated with the text; they illustrate the regional differences described above.

Fig. 2-24. Another wonderful example of a child's wool wheel with unusual reversed legs. Unusual tension device too. The spokes are squared and the overall appearance is somewhat crude. Possibly Canadian. DW=32 ½". *Courtesy of Jeanne Asplundh.* $400-500.

Fig. 2-25. A very nice example of a wool wheel with a rotating barrel tension device, the type of spokes typical of an eastern Pennsylvania wheel, and the little "horns" to hold wool rolags ready for spinning. DW=42". *Courtesy of Jeanne Asplundh.* $500-600.

Fig. 2-26. Wonderful example of an eastern Pennsylvania wool wheel with the distinctive spoke. The maiden finials are very much like those found on documented pieces from the Philadelphia area, now Bucks and Montgomery counties. DW=42 ½". *Courtesy of Jeanne Asplundh.* $700-800.

Fig. 2-27. No picture can do this Berks county wool wheel justice! Pristine chocolate finish with decorative orange bands peeking through. Punched with stars into the table top is the date 1797. If only it were signed, sigh! DW=48 ½". *Courtesy of Jeanne Asplundh.* $1200+

Fig. 2-28. A beautiful wool wheel but hard to pin down where it is from, and it could be anywhere from Connecticut to Ohio. DW=44 ½". *Courtesy of David Pennington.* $600-700.

Fig. 2-30. Signed "C. Porter", this New York area wool wheel is very similar to the work of the Farnham family and E.S. Williams. DW=46 ½". *Courtesy of David Pennington.* $400-500.

Fig. 2-29. A nice example of an "E.S. Williams" wool wheel. Enoch Slossen Williams was a cousin of the famous maker, Joel Farnham, and he was a prolific maker in his own right. Both are covered in the chapter on famous wheel makers. The drive wheel has five grooves in it. Farnham's had eight grooves. DW=46". *Courtesy of Jeanne Asplundh.* $600-650.

Fig. 2-31. Signed "N.D." for Deacon Nathaniel Draper of Enfield, New Hampshire, this wool wheel is the classic New England wheel. DW=45" *Courtesy of David Pennington.* New England Shaker wool wheels, whether signed or not, do not bring premium money because of the large supply, $400-500.

Fig. 2-32. This signed "Thomson" wool wheel (not Shaker) is a New England wheel whose design elements were likely influenced by Shaker wheels sold by the New Hampshire and Maine Shaker communities, which sold thousands over a 90-year period beginning in the 1790s. DW=41". *Courtesy of David Pennington.* $250-350.

Fig. 2-34. This wool wheel is probably Canadian but may be from New England. The tension device is unusual. A very long post on a Minor's head can be moved up and down the deep hole in the spindle post. A wooden thumb screw holds the Minor's head away from the drive wheel. DW=45". *Courtesy of David Pennington.* $250-300.

Fig. 2-35. An unusual tension setup with tension screw set quite high. Overall style looks like New England. DW=45" *Courtesy of Jeanne Asplundh.* $250-350.

Fig. 2-33. This signed "Thomson" combination bobbin and quill winder would have been used in a weaver's studio, so most such winders are rarer, yet less desirable, than wool wheels. This one is a superb example with a beautiful quarter-sawn oak table. Because it has a pointed tip on the iron spindle, it could be used for winding quills for weaver's shuttles as well as for filling bobbins. Theoretically, it could also be small wool wheel, and the absence of a peg on one of the spokes makes its use as a small wool wheel quite possible. DW=27". *Courtesy of David Pennington.* $250-300.

Fig. 2-36. While looking very much like an early Alfred, Maine Shaker wheel with the distinctive tension system, this wheel is different. It has a wooden axle, no horn collar on the spindle post, and the single leg is located between the spindle post and the wheel post. All signed Alfred Shaker examples have metal axles, horn collars, and the single leg is on the end side of the table. DW=43". *Courtesy of James Munsie.* $300-350.

29

Chapter 3
American Saxonies

The saxony is a particular form of flax wheel, and the one people are most familiar with. (In addition to wheels described in the text, a "gallery" of saxonies with descriptions is shown at the end of this chapter.) They normally have three legs, a table, a foot treadle, a bobbin and flyer unit, a drive wheel with a substantial rim (unlike the simple hoop rim of a wool wheel), and usually have a distaff assembly. The key features are that the drive wheel is to the right of the bobbin and flyer unit, and the table is somewhat slanted. Because of the slant, the wheel supports may have supporting posts ending either in the table or the legs. Holes for pegs are found in the tops of the wheel posts, especially the front one, to keep the axle from tipping up as the spinner pushes down on the crank in the rear. The saxony is found throughout northern Europe, but those from the Continent are usually noticeably smaller.

Fig. 3-1 shows the comparison of an American saxony and a European one. The one commonly made in the United States was a variant of what was called the Low Irish or Low Dutch type by continental authors and what we call the American saxony.[1] The bobbin and flyer unit on the American saxony employs a doubled drive cord with the bobbin free to turn on the flyer's shaft. Since the bobbin's whorl is smaller than the flyer's whorl, the bobbin turns more rapidly and draws the thread on as it spins at a rate determined by the relative circumferences of the two pulleys and the tightness of the tension on the drive cord.

Fig. 3-1. A classic American saxony from New England on the right, and a typical European, likely German, saxony on the left. The size differential is typical. *Courtesy of David Pennington.*
New England saxony. $350-450.
European saxony. $150-250.

Fig. 3-2. Fantastic example of a "Picardy" flax wheel in complete, if somewhat "wormy", condition with extra bobbins in the drawer. Found in Connecticut it could come from an early Huguenot family there, but more likely is a souvenir from France. DW=16". *Courtesy of David Pennington.* $700-800.

American saxonies were made throughout the United States in the 18th and 19th century, and we assume the 17th as well. However, since the British were adopting a variety of continental spinning wheel forms in the 17th century, the predominant types of spinning wheels in various colonies reflected the regions from which they came. Some were probably not even saxonies and may have been hand-cranked rather than treadle driven.[2] The wheel in Fig. 3-2 is sometimes referred to as a Picardy wheel after the region in France where they were common. In this arrangement, called a "flyer lead," the bobbin and flyer unit hangs out in front of the maidens and only the flyer is driven by the drive band (see Fig. 1-6). The bobbin is slowed by a retarding string which can be loosened or tightened to give the required speed differential between the bobbin and flyer unit. This bobbin and flyer system is found on very old French-Canadian wheels and on some late 19th century fancy wheels from eastern Europe. There was a determined effort in the late 17th century to introduce the Picardy saxony wheel to Ireland because it was considered a superior wheel for fine flax spinning. Some wheels of this type have both a treadle and a hand-crank, so that the spinner could decide whether greater control was needed or not. In American colonies where French-Huguenot refugees settled in the late 17th century, it would have made sense that they brought or made such wheels. Other types of European wheels are covered in our chapter on European exotic wheels.

Fig. 3-7. Marked "W. Kelia" this saxony is typical of the New Hampshire wheels. Initially we thought this wheel must be Shaker, but extensive searches of all Shaker indexes of names turned up nothing. DW=20". *Courtesy of David Pennington.* $450-550.

Fig. 3-9. This saxony signed "W.W." is a wonderful example of a saxony wheel from the North Shore of Massachusetts. Other signed examples are in the American Textile History Museum. DW=20" *Courtesy of David Pennington.* $500-600.

Fig. 3-8. This wheel represent a group of saxonies found just west of Boston and shows how going only about 20-30 miles can affect turnings noticeably, but the plainness of the New England form is retained. DW=20". *Courtesy of Dave Pennington.* $300-400.

The wheel shown in Fig. 3-7 is marked "W. Kelia," who is not on any Shaker index of names, but his wheel is certainly akin to the Canterbury, New Hampshire Shaker wheel shown in Fig. 3-5. The legs on these two wheels are good examples of what Dave calls the "New Hampshire" leg. Just west of Boston, one finds saxonies like the one shown in Fig. 3-8. The leg has a distinctive single reel turning near the foot. Its otherwise simple turnings place it squarely in the New England tradition. Just north of Boston are found saxonies (Fig. 3-9) with a distinctive decoration on the end grain of the table. There are double horizontal beads above and below the tension screw, and the turnings on spokes and legs are much more intricate. There are several different names and initials associated with this tradition.

34

Connecticut saxonies are much more reminiscent of urban Pennsylvania or the North Shore of Massachusetts than their more spartan New England relatives. Fig. 3-10 shows a nice example marked "Alexander." The turnings on the maiden finials, the black decorative bands, and the diamond patterned chip carvings on the distaff support post are commonly seen on double flyer wheels by known Connecticut spinning wheel makers. But who was this Alexander and exactly where and when did he work? Wheel makers were often turners and turners were also sometimes chair makers. There are over 2000 names of chair makers in Nancy Goyne Evans book, *American Windsor Chairs*, which also includes the names of over 80 spinning wheel makers in the text itself. The editor of *The Spinning Wheel Sleuth* has published four lists of known spinning wheel makers, but alas no information about Alexander has turned up. Since the wheel was collected in Connecticut, Dave might call or write the local historical societies in the towns near where the wheel was collected, if he thought it had not already "traveled" some prior to his purchase of it. So for now, we are simply confident in showing it as an excellent example of the Connecticut saxony.

In a Chapter 11, we take up in greater detail the saxonies of eastern Pennsylvania. Here, however, we show in Fig. 3-11 a classic example of the wonderful wheels from that region. The wonderful turnings and classic foot are evident in this totally complete example marked "J. Klein."

Fig. 3-10. A superb signed "Alexander" example of the Connecticut saxony. However since we know next to nothing about the maker after 30 years of searching, the mark does not materially affect the value of the wheel, but the completeness and decorative touches do. DW=19". *Courtesy of David Pennington.* $550-650.

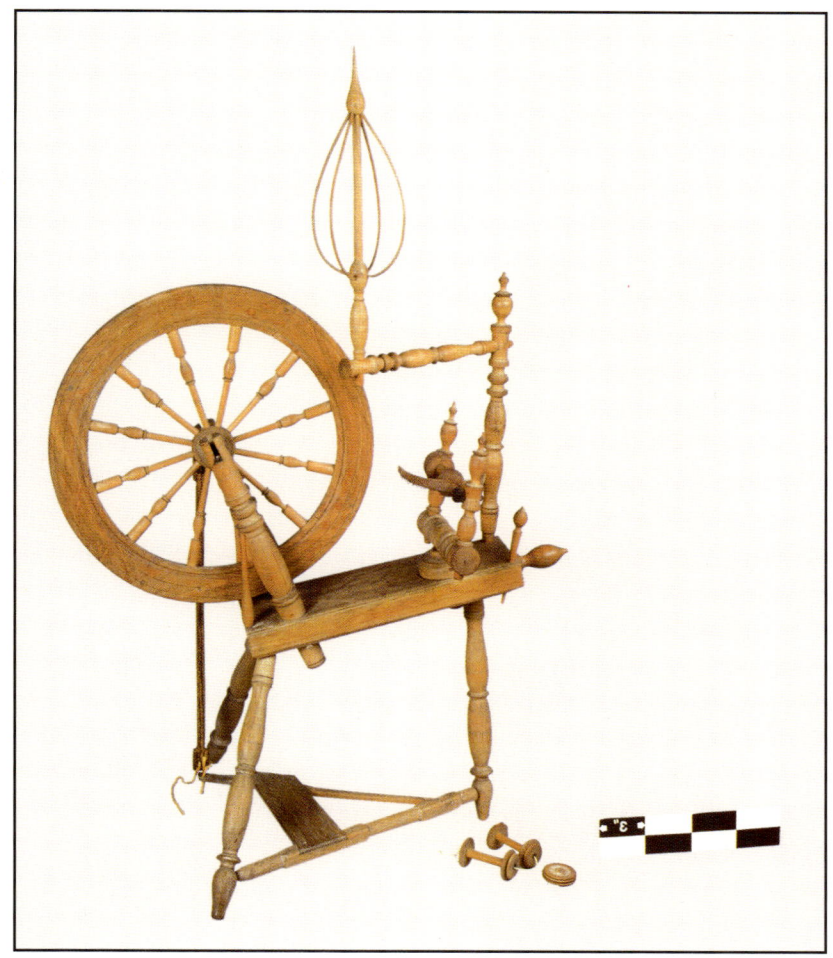

Fig. 3-11. A signed "J. Klein" saxony from eastern Pennsylvania. It is in extraordinary condition with the reeling pin, extra bobbins, and even an extra flyer whorl. DW=19 ½". *Courtesy of Jeanne Asplundh.* $750-850.

Fig. 3-12. A nice example, signed "T. Carter", of the style of saxony found in the western part of New York. The plain spokes, simple legs, and small tipped feet suggest that region. The wheels by the Farnham family, including their cousin, Enoch Slosson Williams, all have these features. DW=20 ½". *Courtesy of David Pennington.* $350-450.

New York state saxonies frequently have a distinctive foot (Fig. 3-12). They are also "leaner" than their New England cousins. So one looks for the combination of features when making an educated guess about where a wheel is from. In upstate New York, like the other regions, there are a variety of finials associated with particular localities and particular makers. Check the chapter on eastern Pennsylvania wheels for some examples of how small pockets of distinctive turning styles and even structural features can develop. Some makers employ more than one finial style, but that is unusual.

In Fig. 3-13 we show a saxony from western Pennsylvania or Ohio. The saxony leg looks like the eastern Pennsylvania type but with a blunt foot. The spoke is a simple bulge with a score mark. Wheels from the Midwest are less often marked with initials and seemed to have "wandered" more from their local regions, so that we have had more trouble pinning them down than wheels from the "old Northeast" and mid-Atlantic regions. Mike has been living in Marietta, Ohio for 25 years and still has only a vague idea about what saxonies in southeastern Ohio look like. Luckily as we noted in the wool wheel chapter, there was a Washington County, Ohio, wool wheel maker, who marked his wheels, but while rumors of signed saxonies by him surface, no one can produce one.[4]

Immigrant communities in the Midwest made distinctive saxonies and left enough examples to confirm that immigrants often made wheels based on styles from their homeland rather than the new country.[5] In 1969 Dave went to an auction of the furnishings of an old pharmacy near Dundee, Michigan. During the Depression its owner had traded goods for spinning wheels from local families. It was an area settled by Germans and sure enough the saxonies there were a simplified German-American form (see Fig. 3-14). In 1970 Dave sold Mike a similar wheel, which he bought near Traverse City, Michigan, from the family of the original owner. Later research suggests that it was the product a Frank Fell, who worked in Mayville, Wisconsin, across Lake Michigan. He worked into the 1920s carrying on the tradition of German-American wheel-making in that region of Wisconsin. He made over 300 wheels a year starting in 1905 when he took over production from the Mayville Furniture Company, which had handled one order for a 1000 spinning wheels in 1900![6]

Fig. 3-13. The kind of saxony that we associate with Ohio but about which we know precious little. A nice decoration, it has little interest to a collector, who wants something more nicely turned or by a known maker. The bobbin and flyer are original and the drive wheel runs true. It is a nice spinner. DW=21". *Courtesy of David Pennington.* Without a mark and missing the distaff pieces $250-300.

Fig. 3-14. Little Germanic saxonies like this one turn up in areas settled by Germans in the 19th century throughout Michigan, Ohio, and Wisconsin. They are also widely available as imports or on Ebay. Some had distaffs and some did not. DW=12 ½". *Courtesy of David Pennington.* $150-250.

Fig. 3-15. This Dutch-American saxony was purchased from a family in Holland, Michigan. Several nearly identical others have been seen there over the years. This wheel has no distaff assembly and was probably used for wool spinning as it has big flyer hooks and a big orifice opening. It could have been used with a free-standing distaff or one held in the lap. It is complete as is. Wheels of this style, but somewhat smaller, are showing up in on-line auctions from Europe. DW=15". *Courtesy of Jeanne Asplundh.* With a solid American provenance like this one, $450-500; the recent imports only $200-250 even with attachments.

Dutch settlers arrived in western Michigan in the 1840s and established communities with the likely names of Holland and Zeeland. Several Dutch style wheels have been found in that area, including the one shown here, which was purchased directly from a Holland family in the 1950s (Fig. 3-15).

Fig. 3-16. Stamped "Paradis" this orange-painted, pine saxony is typical of Canadian saxonies which flooded the antique market in the 1960's and 70's. No distaff pieces as the wheels were designed to spin short-fibered wool. DW=25". *Courtesy of David Pennington.* $300-350.

Fig. 3-17. With a 30+ inch drive wheel and two treadles, this Canadian saxony was definitely meant for production. Their double treadle arrangement makes them something of a novelty in the saxony style. DW=30 ½". *Courtesy of Jeanne Asplundh.* $325-375.

Fig. 3-18. On this early French-Canadian saxony-like wheel, the wheel posts were made by splitting a turned post in half. The tension device is a bobbin lead with a retarding cord slowing down the flyer. This example also came with a spindle so that it could function as a bobbin winder. There never was a distaff. DW=33". *Courtesy of Jeanne Asplundh.* $350-400.

Often we get stories from families about their wheels which just don't add up, so we are very skeptical of oral traditions about wheels unless there is something about the wheel itself that supports the story.

In Quebec, Canada many large saxonies were made into the 1940s.[7] Somewhat crudely finished and then brightly painted in yellow, red, orange, or heavily shellacked, these saxonies (see Fig. 3-16) were designed to spin wool. They have very large orifices and hooks suitable for wool spinning. They do not have distaffs. Some even have double treadles (Fig. 3-17). Their tensioning systems allow the mother-of-all to rotate toward or away from the drive wheels. Various types of clamps hold the mother-of-alls in place. Many rate them quite highly as spinners. They flooded into the United States in the 1960s and continue to be widely available. 18th and 19th century versions were more obviously influenced by the French culture, and the one in Fig. 3-18 is a nice example. It has wheel post support pieces which run diagonally back toward the spinning mechanism, which is referred to as a bobbin lead because the single cord drives the bobbin only. A band running from the crossbar to a small turned peg can be tightened against the flyer whorl and slow down the flyer causing the speed differential needed to allow the spun thread to wind onto the bobbin. Other Canadian saxonies by known makers such as the Youngs family in Nova Scotia are quite beautiful and refined.[8] They have bands of color, which are somewhat reminiscent of saxonies from Berks County, Pennsylvania.

---

[1] John. Horner, *The Linen Trade of Europe during the Spinning-wheel Period*, (Belfast: M'Caw, Stevenson and Orr, 1920): 18-19. Horner makes the point that Dutch wheels had smaller drive wheels which are much thicker than the "low Irish" system.

[2] Ibid., 30-31. Irish saxonies and Dutch saxonies did have treadles. Louis Crommelin, a French Huguenot refugee, who was made Overseer of the Royal Manufacture of Linen in Ireland, complained in 1705 about the treadles on wheels in his discussion of his work in Ireland. When saxonies in the American colonies began to use treadles routinely is a matter of some debate, but a good guess is around 1730. Laurel Thatcher Ulrich is very careful in her discussion of the matter in *The Age of Homespun: Objects and Stories in the Creation of an American Myth*, (New York: Alfred A. Knopf, 1991): 101-102.

[3] Nancy G. Evans, *American Windsor Furniture: Specialized Forms*, (New York: Hudson Hills Press, 1997): 219-221.

[4] William Reynolds, "Some More Thoughts on the Vincent Spinning Wheel," *The Association of Ohio Long Rifle Collectors*, Vol. IV, Nancy G. No. 1 (February, 1982): 14-15.

[5] Victor L Hilts, and Patricia A. Hilts, "Not for Pioneers Only: The Story of Wisconsin's Spinning Wheels," *Wisconsin Magazine of History* (Autumn 1982): 2, 18-19. Laura Gustafson, "Scandinavian Spinning Wheels," *The Spinning Wheel Sleuth* 14 (October 1996): 8-9 is really about Scandinavian types of wheels that Ms. Gustafson regularly sees in Minnesota.

[6] Ibid., 21-22.

[7] Judith Buxton-Keenlyside, *Selected Canadian Spinning Wheels in Perspective: An Analytical Approach*, (Ottawa: National Museums of Canada, 1980): 276.

[8] Ibid., 154-158.

Gallery of American Saxonies

The flax wheels in this Gallery are not inferior to the ones illustrated with the text; they illustrate the regional differences described above.

Fig. 3-19. This Dutch style saxony was found in Michigan many years ago and may well have been made by Dutch settlers near Holland, Michigan, but unlike the wheel in Fig. 3-15 there are no other nearly identical others to tie it to the region. Only wood analysis will tell for sure. The bobbin rack for plying yarns is a wonderful plus. Many Dutch wheels are showing up on eBay. DW=16". *Courtesy of David Pennington* $350-400.

Fig. 3-20. Another Dutch style saxony collected in Michigan many years ago, but with no known direct tie to Dutch settlers there. It is so classically Dutch that we suspect it is, but again only wood analysis would pin down where it was made. DW=15". *Courtesy of David Pennington*. $250-300.

Fig. 3-21. We lavish attention on this wheel because it is an extraordinary example of the American saxony, but unfortunately we know little about its specifics. The turnings do not fit with a locale we are familiar with, but the existence of a reeling pin and wheel post suggests eastern Pennsylvania (see Chapter 11). The wheel has a wonderful early finish and is totally complete, but it is unsigned. *Courtesy of Jeanne Asplundh*. $700-750.

Fig. 3-22. Wonderful turnings on the maiden finials and distaff cross piece. They are earlier in style than we often see, and we suspect this wheel is an 18th century Philadelphia piece. *Courtesy of Jeanne Asplundh.*

Fig. 3-24. This eastern Pennsylvania saxony has both a flax distaff and a tow distaff. Interestingly, there are 9 flyer teeth on one flyer arm and 13 on the other, which would be handy if the wheel were intended to be used for flax and tow. Tow is a heavier thread. It has a script "L.M." picked into the table. A superb wheel. DW=20". *Courtesy of Jeanne Asplundh.* $600-650.

Fig. 3-23. This close up of the mother-of-all shows a reeling pin which is also used to keep the tension screw in place. Many wheels simply have a peg driven in here which must be driven out from the bottom if you want to remove the mother-of-all and tension screw. Reeling pins were a feature on many eastern Pennsylvania wheels. *Courtesy of Jeanne Asplundh.*

Chapter 4
"Unusual" Flax Wheels

Fig. 4-1. The treadle on this saxony is reversed, i.e. "left-footed". Since some wheels show from their wear marks that they were treadled with both feet, this design feature is not for the left-footed but is to allow for a more upright stance which saves floor space. The turnings suggest a Midwestern origin, perhaps Ohio where it was found. DW=20". *Courtesy of Dave Pennington.* $550-600.

A variety of unusual flax wheel styles are rarely encountered and seem to be American in origin, or in execution of form. They all have a single treadle, a single wheel, and a single bobbin and flyer. Some of them may well be patented, but because they are not marked and because the patent office files burned in 1836, there is a real problem knowing for sure. Occasionally, odd wheels are marked, and we have a survey list of patentees, dates, places that gives us some confidence in assigning a wheel to a pre-1836 patent. Those wheels we cover in our chapter on patents or in the chapter on famous makers. The ones here are interesting and "unusual." They are unusual in being odd types and scarce. We cover double flyer wheels; double treadle, double wheel flax wheels; and Irish castle wheels in separate chapters because they were popular enough to leave dozens of examples for study.

The "left-footed saxony" appears in several regions. The first one shown here (Fig. 4-1) could be from eastern Pennsylvania given its "honey-dipper" turnings on the finials of the maidens and the distaff holder, but the spoke and the wheel post look more Midwestern. Notice that the treadle arrangement is backward, which we laughingly called left-footed. But as wear marks on treadles clearly show, most wheels were treadled with both feet, and the reason for this configuration of wheel is to save space. By having the treadle arrangement this way, the drive wheel does not extend over the end of the table, but is located more directly over it. Of course, that means that the flyer arms must be positioned so that they do not hit the drive wheel. The second example (Fig. 4-2) is interesting because it is definitely eastern Pennsylvania and probably Chester County. It is dated 1822, so we can see that by then space is becoming an issue. Left-footed saxonies with New England turnings have also been seen. Several virtually identical examples to the one shown in Fig. 4-3 have western Pennsylvania associations.

Fig. 4-2. This signed Pennsylvania wheel is extensively restored but well worth the effort and expense. It is a "left-footed" version of a group of wheels from eastern Pennsylvania. These wheels, examined in detail in Chapter 11, have a number of distinctive features, most notably, drive wheels constructed of four irregularly shaped pieces. The wheel is signed, dated and numbered in a manner common to Bucks County makers and some Berks County makers. It is likely the work of William J. Major of Chester Springs, Pennsylvania, in Chester County. He advertised as a spinning wheel maker in 1829 in Chester Springs, Pennsylvania and was listed as a turner in Chester City, Pennsylvania, in an 1842 census. This wheel is dated 1822 and is marked "W+M". The mark is picked with an awl rather than stamped, so the "+" is likely a "J". A decorative "sine wave" motif using punched stars on the front edge of the table pushes this wheel into the highly collectible category given its being a "left-footed" saxony, having a date, probable maker, and established region of origin. If complete this wheel would be a $1000+ saxony. DW=17 ½" *Courtesy of Michael Taylor*. $700 as restored.

Fig. 4-3. The profile of this "left-footed" emphasizes its compactness. The movement from horizontal to vertical is accentuated by tall wheel supports. Four examples of this wheel maker's work have been spotted. All of them were in various ways associated with western Pennsylvania. None are signed. This example, while relatively complete, had been "skinned", i.e. the old finish removed, a definite no-no. Patient oiling over the last 25 years has given it an acceptable look. DW=19 ½". *Courtesy of David Pennington*. $700.

The wheel in Fig. 4-4 has four legs and is so upright that it can no longer be called a saxony, although the position of the tension screw suggests that it evolved from one. Here again saving space would seem to be the point of the design. This one is in black and gold paint, which was applied after the wheel was taken out of use. It is not common to see such painted wheels in eastern Pennsylvania, but even more rarely was the paint applied when the wheel was made. Usually close observation of wear marks on the treadle and the flyer arms will reveal if the paint was original. However, the paint may well be late 19th century, so removing it is not a good idea.

Fig. 4-4. Although very similar to Fig. 4-3, this wheel has moved to a total vertical orientation. The treadle no longer has the diagonal piece and now consists of treadle bar and treadle. Most importantly, it now has four legs instead of three. It can no longer be called a saxony in any sense. It is painted in old black and gold paint. When it was painted is hard to say. So far, this is the only known example of this style. Its turnings strongly suggest a Pennsylvania origin. The back maiden has no finial. DW=20 ½". *Courtesy of David Pennington*. $900-1000.

bobbin and flyer unit. It has the name "W. H. Logan" painted on it, but we do not know if that was the wheel maker, the painter, or the owner's name. (Wheels were commonly stained and even occasionally "mahoganied," but they were seldom painted for decoration. Sometimes wheel hubs or other parts might be painted to keep them from cracking later.)

The wheel in Fig. 4-6 is another example of a double flyer form adapted for use as a single flyer, which is metal. Double flyers of this type are not uncommon in Connecticut, but this single flyer example is very uncommon. Some of the double flyer models have metal flyers, a feature also seen on European wheels. Baines and others have pointed out that in Great Britain, the term "castle" wheel is reserved for wheels, such as this one, on which the drive wheel is above the bobbin and flyer unit. The threaded tension rod on this example is frequently found on New England upright double flyer wheels, so we are confident that this one is American.

Fig. 4-5. With its early paint and stenciling this extremely unusual wheel is folk art as well as textile tool. The name "W.H. Logan" is painted on, but whether he is maker, owner, or painter is unclear. The form is loosely based on a type of double flyer wheel commonly found in Connecticut. This one, collected in Pennsylvania, may well be from Pennsylvania. DW=17 ½". *Courtesy of David Pennington.* $900-1000.

The wheel in Fig. 4-5 probably was painted and stenciled very early on. The style is loosely based on a popular form of double flyer wheel from Connecticut, but a close look at the tension system reveals that the whole upper section moves up and down rather than just the

Fig. 4-6. An interesting example of what the English call a "castle" wheel. This form is more commonly seen in New England as a double flyer. The threaded tension rod and turnings suggest an American wheel, and the metal flyer is seen on both sides of the Atlantic. DW=19 1/2". *Courtesy of David Pennington.* $700-750.

The next wheel Fig. 4-7 is a small upright wheel marked "MA.TT CALHOON" with the drive wheel below the bobbin and flyer. We are reasonably sure that, while it has a very European look, it is American in origin. A very American saxony with the same stamped mark is known. Having the maidens pointing downward is quite unusual as are the wheel post support pieces.

Another remarkably similar one has the marks "NG No. 644" carved in the front side of the table (Figs. 4-8 and 4-9). It was found in Maryland, but it is very similar to the Calhoon wheel, which is almost certainly from eastern Ohio. Undoubtedly, small upright wheels were made in 19th century immigrant communities in the United States, as with Ukrainians in Manitoba, Canada, but the small upright type does not seem have been popular in the United States.

Fig. 4-8. Amazing similarity to the preceding wheel but with "N.G. No. 644" carefully picked in the table. Since there is a number, it is likely a maker's mark. DW=20". *Courtesy of Jeanne Asplundh.* $500-550.

Fig. 4-7. This wheel certainly looks European, but the stamp "MA.TT CALHOON" has been seen on very American looking saxonies in eastern Ohio. In A *Pictorial Guide to American Spinning Wheels* we referred to wheels like this one as "castle" wheels or parlor wheels. British collectors and authors use castle to refer only to spinning wheels with drive wheels located above the bobbin and flyers. Parlor wheel can refer to any small fancy wheel with inlays or decorative bone or ivory detailing. In spite of the relative rarity of American examples of this type of wheel, it is not particularly desirable. If we knew more about Matt Calhoon, that would change quickly. DW=20". *Courtesy of David Pennington.* $500-550.

Fig. 4-9. Close up of the mark in Fig. 4-8 which shows how initials may be "picked" into a table. *Courtesy of Jeanne Asplundh.*

In Fig. 4-10 we see an adaptation of the Connecticut chair wheel (see Fig. 6-1). The chair frame design is clear, but there is a single drive wheel and single treadle arrangement. That would save a little time and money in the making and perhaps even enhance the spinning characteristics as there are various problems with the designs that employ two wheels. Getting the belt or cord tension between the two drive wheels just right is sometimes a problem. The double treadles had some advantages for the spinner as she stopped and started spinning, but not enough to warrant the extra expense.

Here in Fig. 4-11 is a double treadle wheel with a single drive wheel. With its split table it somewhat resembles a double treadle, double wheel marked "B.B." from Joan Cummer's collection. This one was collected in North Adams, Massachusetts and is likely a New England wheel. While we give Alpheus Webster of Green County, New York credit for developing the double treadle because of a signed example and two early patents, this wheel is a fine example of an independent execution of this combination of features.

Fig. 4-10. Here is a "chair wheel" with a single large drive wheel rather than the common double wheel type. It has a single treadle instead of the common double treadle. Unmarked as is commonly the case with chair wheels, it would be easier and cheaper to make, but we don't know if this one preceded or succeeded its more famous cousin. Definitely rare, but not very attractive. DW=20 ½". *Courtesy of David Pennington.* $700-800.

Fig. 4-11. With a double treadle but only a single drive wheel, this wheel is quite unusual. It does resemble a wheel from Joan Cummer's collection that has two wheels. DW=22 1/4". *Courtesy of David Pennington.* $750-850.

Fig. 4-12. Examples of a "J. Farnham" (left) and an "E.S. Williams" (right). Farnham and Enoch Slosson Williams were cousins. Joel Farnham also trained his sons Joel Jr. and Frederick Augustus in the trade and they also stamped these wheels with their marks. Farnham DW=20". Williams DW=21". *Courtesy of David Pennington.* "J. Farnham" example $850-900. "E.S. Williams" example $750-800.

In Fig. 4-12 we show a type of four-legged wheel made in quantities by Joel Farnham, his sons, and his cousin Enoch Slosson Williams near Owego, New York in the 1820s and later. Every serious wheel collection should have an example of this type. Other makers, like "L. Brown" and Farnham's sons, copied this style which must have been popular along the New York-Pennsylvania border. It is very interesting to compare this style of wheel with that of the traditional Swedish/Norwegian horizontal flax wheel (Fig. 4-13). The most striking visual similarity is the threaded horizontal rod. On the Farnham wheel this threaded rod is the tension device. On the northern European flax wheel, there are two such rods and they are used to align the drive wheel and the bobbin and flyer units. The tension device on this wheel is a traditional threaded rod running through the table with a moveable mother-of-all as on a saxony.

Fig. 4-13. A great example of a Scandinavian type of wheel frequently encountered in Wisconsin, Iowa and Minnesota where it was made in the latter part of the 19th century by Swedish immigrants. They are commonly painted; this one is black with gold bands. We put it here to show its structural similarities to the Farnham clan's work. The two threaded screws are to align the drive wheel with the bobbin whorls and not for tensioning as in the case of the Farnham type. Only wood analysis can determine if it was made in America. DW=23 ½". *Courtesy of Michael Taylor*. $500-600.

Another odd type from upstate New York is seen in Fig. 4-14. The open layout of this type with no main table is interesting. The tension device here is one of several employed; this one uses a threaded rod. Others use a pressure nut similar to that found on most Connecticut chair wheels (Fig. 4-15). Most of the drive wheels in this group are constructed more like the drive wheels on wool wheels with a thin lap of wood nailed together rather than the sturdy, thick-rimmed wheels typical of saxonies. Such drive wheels do not have much "carry" (momentum) when treadled, and the spinner has to work harder to keep the wheel going (Fig. 4-16).

Fig. 4-15. A different example of the western New York wheel with a nut and thread system similar to that employed on many chair wheels. DW=23 ½". *Courtesy of Jeanne Asplundh.* $700-800.

Fig. 4-14. This style appears to have been popular somewhere in western New York, but so far the evidence is tantalizing and mainly based on where a few have been collected. No signed examples are known, but two distinct tension systems were employed. The crank is "internal" i.e. inside the back wheel support post. DW=24". *Courtesy of David Pennington.* $700-800.

Fig. 4-16. Another example of the very rare and desirable upstate New York wheels detailed in Figs. 4-14 and 4-15. This one has a new bobbin and flyer, treadle and distaff assembly staff, done by a highly skilled restorer that would cost $250-350. Unless an identical wheel is the basis for the restoration, it will substantially detract from the wheel's market value. DW=24 ½". *Courtesy of David Pennington.* $450-500.

Fig. 4-17. Shown here with its adapted distaff arrangement, this internal crank wheel is spectacular. The black paint is original and shows the expected wear patterns. It came out of an attic in Berlin, New York with old spun linen thread on the bobbin and a bundle of flax ready to be spun; truly "attic-found". Not only its general style, but details in the way its flyer metal was lathe-turned rather than forged and the horn bushing inside the front leather show its close relation to the wheel in Fig. 4-18. DW=19". *Courtesy of Michael Taylor.* $900-1000.

One of our favorite wheels is shown in Fig. 4-17 and it is closely related to the type shown in Figs. 4-14, 4-15 and 4-16 as well as Fig. 4-18. Its most distinctive feature is the tall upright post in the center of the structure. Luckily, the best feature of the wheel was preserved when its seller insisted that Bill Leinbach, the famous collector/dealer, take the flat board piece because "it is also black like most of this wheel." Bill and Mike fiddled with the piece for hours before it dawned on them - slow learners that they are – that the board fit on the tapered shaft of the tall upright and rested on it and the front wheel post perfectly. That provided the spinner with a "table" and also a place for the distaff. This "table," while quite old and obviously used with this wheel, is a later adaptation. The original distaff was like an old lantern with a hole in the base that was lowered down over the top and its height was adjusted by putting a wooden peg in a hole in the tapered rod.[1] The wheel in Fig. 4-17 is mostly painted black with only the maidens, wheel spokes, and the bobbin and flyer not painted black. Since the treadle and other wear occurred after the black paint was applied, the paint is early, if not original. It was found in an attic in Berlin, New York.[2]

The wheel in Figs. 4-18 and 4-19 is a "killer." An unusual design, it is stamped "W. Fancher" on the top of the square post housing the crosspiece for the maidens. The ample use of tiger-striped maple and the finish used to highlight it suggest a furniture maker. The placement of the stamp is common among chair makers. It has horn bearings instead of leathers to hold the bobbin and flyer. It is the perfect example of understated elegance in a spinning wheel. The tension device is similar to that commonly employed on French Canadian wheels, but the execution of the design with the beautiful little head on the "thumbscrew" is from another tradition as is the internal crank. In some ways it is similar in overall design to the wheels in Figs. 4-14, 4-15 and 4-16, but the drive wheel is much more substantial. Hopefully some descendant of W. Fancher will surface and tell us more about the where and when of this wheel. The only other example of one, also signed by W. Fancher, was described as being in old red paint, which seems unlikely to have been original. The one shown belonged to an early west coast collector and spinner named Irma Green from South Pasadena, California. She found it in Pasadena and claimed to have seen its twin there also. Its simple design, use of figured maple, and odd tensioning system suggest a northern New York or possibly Canadian point of origin. A very plain wool wheel with a rotating barrel tension device with a single post similar to that shown in Fig. 2-5 is also signed "W. Fancher" which would also suggest this region as its point of origin. The black wheel in Fig. 4-17 and the Fancher wheel have the same unusual type of bobbin and flyer suggesting they were made nearby one another in central New York.

Fig. 4-18. With beautiful tiger-striping, untouched finish, exceptional details and a unique design, this "W. Fancher" stamped wheel is one of a handful of truly exceptional wheels along with the "John Burley" wheel, the "A. Webster" signed single drive wheel model, the Guilford solid-wheeled chair wheel in red paint, and the signed "S. Henry," four-legged, Irish castle wheel. The flyer metal is lathe-turned rather than hand forged. There is a horn bearing to hold the front end of the flyer instead of a leather. There are threaded bearings for the drive wheel which can be adjusted to align the drive wheel with the bobbin and flyer. Only the relatively primitive tensioning system is a puzzle. The mother-of-all rotates and a threaded screw holds it in place much like the later Canadian saxonies. DW=21". *Courtesy of Michael Taylor*. $1500-1800.

Fig. 4-19. An end view of the "W. Fancher" wheel. *Courtesy of Michael Taylor.*

Another flax wheel type without a table is shown in Fig. 4-20. It is similar to a saxony in shape, but it has no table and the tension screw runs through the front leg. It is a very original style. The turning on the foot suggests New York State, and the three we know of were all from central New York.[3]

One group of spinning wheels is referred to as "tow" wheels. It was thought that their extraordinarily large orifices were for spinning "tow," which is the term for the short flax fibers not suitable for fine linen. Tow, however, could be quite fine as any linen with under 40 threads per inch is technically "tow." Obviously, flax wheels can handle threads suitable for tow cloth at 20 ends per inch. Perhaps, tow wheels were used to spin and ply tow for cord (Fig. 4-21).

Fig. 4-20. Likely from upstate New York, three of these very unusual wheels have surfaced. There is no table. The turnings are quite nice. The tension system uses a rotating mother-of-all with threaded rod, an uncommon combination. The drive wheel style and foot turning are common in the Owego, New York area. DW=21 ½". *Courtesy of David Pennington*. $600-700.

Fig. 4-21. A "tow" wheel is the name of this type of wheel with a giant orifice. The unusual "lollipop" decorations would have increased the drive wheel's inertia and made it easier to treadle. It is thought that these wheels were used to spin the short flax fibers unsuitable for fine linens. Some think it was used to spin coarse fibers for rope. DW=10". *Courtesy of Jeanne Asplundh*. $400-450.

Small vertical wheels of the sort shown in Fig. 4-22 are quite common throughout Europe, but this one from Great Britain, mostly in metal, is quite unusual.

Fig. 4-22. One of only two mostly metal parlor wheels known. The other is in a museum in Edinburgh, Scotland. This is a form of vertical flax wheel is commonly found in Europe in all quality levels. DW=18". *Courtesy of Jeanne Asplundh.* $2300-2500.

---

[1] Pam Mawhiney wrote about an almost identical wheel in *The Spinning Wheel Sleuth* and included a picture. It was the picture which helped us realize what the original distaff looked like. Interestingly enough, the drive wheel on Mawhiney's example has a thin rim like the wheel in Fig. 4-14. Pam Mawhiney, "A Wheel with a Bentwood Drive Wheel," *The Spinning Wheel Sleuth*, 18 (April 1997): 8.

[2] Virginia D. Parslow, "Spinning Wheels," *Handweaver and Craftsman*, (Spring, 1956): 22 shows a very interesting wheel which is halfway between the wheels shown in Figs. 12 & 14. In 1956 the wheel was in the collection of the Farmer's Museum in Cooperstown, N.Y. From the small variations on the open frame types shown in Figs. 4/12- 4/15 it seems likely that in the area near Berlin, New York this type of flax wheel with no table was popular for some reason.

[3] Bill Ralph, the long-time collector/restorer and founder of the Home Textile Tool Museum, found one in Milan, Pennsylvania near the New York border and thought the wheel quite unusual. "Inquiry," *Spinning Wheel Sleuth*, 4 (January, 1994): 3.

Chapter 5
American-Irish Castle Wheels

John Horner, a spinning wheel collector and writer from the early 20th century, suggested that the Irish castle wheel (Fig. 5-1) was the earliest form of Irish spinning wheel.[1] It was being used in the Ulster area into the 20th century. Horner suggested that they were unique to Ireland, and principally found in counties Antrim and Donegal, but wheels very closely based on them were being made in Lancaster County, Pennsylvania from the 1820s into the 1860s and versions of it were definitely there in 1806. Additionally, another version was being made in New England in the same period. Who brought the style to these regions and how early is the puzzle.

The wheel shown in Fig. 5-2 is actually from the Landisville area in Lancaster County, Pennsylvania, but the spoke turnings, tension device, system for removing the bobbin and flyer unit, and leg turnings are identical to photos and drawings of those from the Ulster, Ireland area. It is one of a number collected near Landisville, but none are marked. Normally, but not always, the metal teeth on the flyer arms are on the right arm rather than the left (as one looks down on the arms).

Left:
Fig. 5-1. An Irish castle wheel, probably imported from Ireland, this example has worm holes. The Irish examples are somewhat heavier and less graceful than their Lancaster County, Pennsylvania counterparts. DW=17 ½". *Courtesy of David Pennington.*
$400-500.

Right:
Fig. 5-2. A super example of what Bill Leinbach, the noted Pennsylvania wheel collector, refers to as Landisville Irish castle wheels. He found several, probably by the same maker, at house sales in the Landisville, Pennsylvania area. This one has two short posts for holding the distaff assembly consisting of a horizontal arm and distaff. They are both old and have the same finish as the rest of the wheel. The distaff is of a type found in eastern Pennsylvania, and it fits the horizontal arm. The flyer teeth are on the top right side which is unusual except in Lancaster County. DW=17". *Courtesy of James Munsie.* $1200-1500.

Chapter 6
Double Treadle, Double Wheel Flax Wheels

Fig. 6-1. This is the "classic" Connecticut chair wheel model of which we have seen a dozen or more. It does look like it was made in a chair frame. DW=14". *Courtesy of David Pennington.* $600-700.

In the late 1960s, the wheel we wanted most was the "Connecticut chair wheel" (Fig. 6-1). It was an American form, most unusual looking, and ingenious. It had attracted early collector/authors, like Wallace Nutting, and two of them were shown in Serge Daniloff's early work on rare wheels.[1] It is not as attractive as an Irish castle wheel, but it has a simple elegance of its own. Not much has changed in our appreciation of it. Other collectors we have met over the years have had the same reaction. Irish castles may be more beautiful, but nothing gives a collector a "rush" like walking into a shop and seeing a chair wheel. Its two most prominent features from a technology standpoint are the two treadles and the extra wheel, hence the generic descriptor: double treadle, double wheel flax wheel. Chair wheels are only one version of this type. Some writers probably influenced by Serge Daniloff's article have suggested a mid 18th century origin for such wheels, but double wheel flax wheels almost certainly are an early 19th century development. In addition to the likely connection to the Minor's head as a mechanical improvement, the turnings on these wheels are appropriate to the 1810-1830 period.

There is little solid information about the origin of the double treadle, double wheel, flax wheel. The second wheel, actually an accelerating wheel, makes it a close relative of Amos Miner's (his name is misspelled on the labels as A. Minor) famous accelerated head for a wool wheel, which was patented in 1810.[2] We think that this very successful improvement to the wool wheel head was picked up by a number of makers, who then applied the principle to the flax wheel. Unfortunately, the Patent Office fire of 1836 wiped out one source of evidence, and only a few signed examples of accelerated flax wheels have survived.

We have pieced together evidence and piled on some surmises to come up with the following likely chain of developments. Alpheus Webster of Greene County, New York, applied for and received two patents for spinning wheels, one in 1810 and one in 1812.[3] Two very different types of flax wheels with the mark "A. Webster" have been found. One is a vertical wheel (Fig. 6-2) with the drive wheel above and two treadles below. Fig. 6-3 shows how the axle crank was modified to accept two footmen. In this case the crank is "external," i.e. outside the wheel posts. The other type (Fig. 6-4) has a similar base with two treadles, but has an accelerating wheel, very close in size to a Minor's head. (A Minor's head accelerating wheel is about five inches in diameter, depending on which firm made it, and Webster's is about six and a half inches.) It seems likely that Webster used the double treadle concept initially around 1810, and then having seen Minor's heads he incorporated the double wheel configuration in his 1812 patent wheel.[4] In 1811 Miner's invention was being widely pirated.[5] Fig. 6-5 shows another change, the "internal" axle crank arrangement, which becomes the norm for double treadle wheels.

Fig. 6-2. One of two double treadle wheels with a single, bent rim drive wheel marked "A. Webster". Alpheus Webster holds two patents, one in 1810 and one in 1812. We believe that this wheel is the 1810 patent version and that after seeing a Minor's head with an accelerating wheel Webster took out a patent on a double treadle, double model in 1812. This one is in wonderful condition and of great historical significance. The other one is at the Farmer's Museum in Cooperstown, New York. DW=24". *Courtesy of David Pennington.* $2500-3000.

Fig. 6-3. A close up of the previous "A. Webster" wheel showing the external axle crank for the two footmen. It also shows the external brass bushing which is unique. *Courtesy of David Pennington.*

Fig. 6-5. Close up of the "internal" axle crank devised by Alpheus Webster for his double treadle, double wheel flax wheel. This arrangement became the norm. *Courtesy of David Pennington.*

Fig. 6-4. This double treadle, double wheel is an unsigned example of the other "A. Webster" wheel. Two signed examples are known, one at the American Textile History Museum. The distinctive features of the signed examples are the metal rim with wooden spokes, the turnings, and a fascinating tension screw. Four or five unsigned examples have been spotted with identical features. Our hypothesis is that those sold by retailers at a distance from the maker were marked by the maker, both as a kind of warranty and for promotional purposes. DW=14". *Courtesy of David Pennington.* $1100-1200.

The drive wheels on the two examples signed "A. Webster" have wooden spokes but a metal rim. A drive belt, rather than a cord, supplies the power to the accelerating wheel which has two grooves to receive the traditional doubled cord for driving the bobbin and flyer unit. The accelerating wheel is painted black, like a Minor's head, to prevent the solid wheel from cracking as the wheel dried out. Webster's double wheel type became relatively popular with many signed and unsigned examples, some even with two bobbin and flyer units. Some makers copied it closely. Beardsley Sanford of Fergusonville, New York copied it directly, signed them "B. Sanford" and changed very little (Fig. 6-6).[6] Either Webster or someone else (although we strongly suspect it was Webster) went to an all-metal, cast, drive wheel with four spokes and simplified the threaded tension screw arrangement (Figs. 6-7 and 6-8). The tension screw on the earlier Alpheus Webster wheels with the metal rims and wooden spokes has to be seen in person to be fully appreciated. The one piece handle and threaded screw have been hollowed (drilled) out. The simplified versions do not require this bit of wood-working dexterity.

Fig. 6-6. A phenomenal example signed "B. Sanford". Beardsley Sanford made wheels in Fergusonville, New York. It is not far from where Alpheus Webster was working. DW=14". *Courtesy of Jeanne Asplundh.* $1200-1300.

63

Fig. 6-7. While this wheel is very similar to the preceding example, the drive wheel is cast iron with four spokes instead of six. The unusual and difficult tension screw has also been redesigned. None of the all metal wheel examples are signed, but we believe that they represent later production developments of Webster because of the similarities in turnings. We show one example of this wheel (Fig. 6-29) with two bobbin and flyers. DW=14 ½". *Courtesy of Jeanne Asplundh.* $800-850.

Below:
Fig. 6-8. Close up of the preceding wheel showing distinctive Webster turnings and the later cast iron rim and tension screw. *Courtesy of Jeanne Asplundh.*

Other makers used all-metal drive wheels or spoked accelerating wheels (Fig. 6-9). Enough different versions have survived for us to say that it was a relatively successful type in areas in New York. One inventive soul took the wheel type and used it as a platform for the most interesting variation on a bobbin and flyer we have ever seen (Fig. 6-10).

We have seen six of these double treadle, double wheels which are virtually identical in turnings and in the very unusual way in which the threaded rod for the tension device and the drive wheel were made. Two of these are marked "A. Webster," but we believe all six were made by him and that he marked those he put out for sale with retail merchants. The double flyer wheel shown has identical turnings, but a one-piece cast iron wheel and a different way of making the tension screw. Others with a single bobbin and flyer are also known. Are these made by Webster? Dave Pennington quite plausibly argues in the affirmative. Since Webster was an inventor, it would make sense that he would improve his basic design. Casting the wheel in one piece is more efficient if one intends to make them in quantity, and the new tension screw was much easier to produce. So in Webster's case we probably see someone not only patenting new wheel forms, but also tinkering with features to make production easier.

Left:
Fig. 6-9. An unsigned version with a spoked accelerating wheel. A few examples with spoked accelerating wheels show up, but none are signed. They testify to the popularity of this style near the Hudson River in the early 19th century. DW=15 ½". *Courtesy of Jeanne Asplundh.* $750-850.

Fig. 6-10. On the left is the disassembled bobbin and flyer from a typical single flyer wheel like Fig. 6-4. Next to it is an atypical disassembled bobbin and flyer from the wheel in Fig. 6-7. A dramatic innovation in the bobbin and flyer mechanism, it never caught on as it gave no obvious advantages and was more costly to make. *Courtesy of Jeanne Asplundh.* Innovative version of bobbin and flyer $125; standard version $80-100.

Fig. 6-11. Extensively restored with new bobbin and flyer, new drive wheel, and new distaff pieces and a poor finish, this wheel's main claim to value is its being marked. "J. Miles" is the only complete name known on a Connecticut chair wheel, and his name also appears on a superb double flyer wheel, but we know nothing more about him. Two or three of his signed chair wheels are around, and this one is the best of them. It is definitely a collector's piece. DW=14". *Courtesy of Michael Taylor*. $500-550.

From upstate New York the double treadle, double wheel idea spread east and west. In nearby Connecticut someone picked up the idea and developed the Connecticut chair wheel with which we started our discussion of double treadle, double wheel flax wheels. "J. Miles," who signed double flyer wheels, also signed some chair wheels (Figs. 6-11 and 6-12). Unfortunately, no one else gave more than their initials, and no J. Miles appears in the patent records. Connecticut chair wheels are best known in the form shown in Fig. 6-1 with plain round posts and the threaded nut tension device on the left front post. Why they are called chair wheels is evident as it looks as if some ingenious chair maker had simply put two wheels in a chair frame. A distinctive and unfortunate feature of this particular form of Connecticut chair wheel is that there is no means of adjusting the tension between the drive wheels, so there is a slippage problem.

Fig. 6-12. The "J. Miles" mark. *Courtesy of Michael Taylor*.

Around Guilford, Connecticut, two related versions of the chair wheel developed in which this type reached its highest state of development (Figs. 6-13 and 6-14). The spoked version (Fig. 6-13) has an "E.L." stamp on the back left post as a chair maker might mark his chair. Four of these "E.L. stamped wheels are known to the authors, and there is a very similar one marked "G.W." on the top of the front legs (Fig. 6-16). Surprisingly, it is about the unsigned solid wheel versions that we think we know the most. In 1904 a Joseph Wyatt donated one of the solid wheel Guilford chair wheels to what became the Henry Whitfield State Museum in Guilford, Connecticut.

The accession card notes that it was "invented by Joshua Goldsmith and preferred by many of his townswomen to the older pattern." Goldsmith died in 1838, so he was alive and working in the right time period to be making chair wheels.[7] Five of these solid wheels are in museums in Guilford strengthening the tie between these wheels and Guilford. The top solid wheel has holes partially drilled in the back. We once thought lead weights were inserted to give the wheel more momentum, but we now are just plain puzzled about their purpose. The augur used to drill the partial hole was gimlet-pointed, which is a 19th century feature and not an 18th century one (Fig. 6-15). The "E.L." stamp may have belonged to Edmund or Eli Leete who lived in Guilford in the right time period and had carpenters' tools but no lathes.

Below:
Fig. 6-14. The solid wheel version of the Guilford, Connecticut chair wheel. There is good reason to believe that Joshua Goldsmith was the maker of these wheels. Solid wheels are easier to make but may split if the wood is not well seasoned or painted. The finish on this wheel is excellent, but the distaff assembly has been replaced. DW=14". *Courtesy of David Pennington.* $750-850.

Fig. 6-15. Rear view of the solid double showing the distinctive marks from a gimlet pointed augur. Frequently found on the top wheel, the purpose for them is unknown. *Courtesy of David Pennington.*

Fig. 6-13. Absolutely wonderful example of the fully developed Connecticut chair wheel form probably from Guilford, Connecticut. The top wheel may be raised or lowered to adjust the tension on the drive belt. In the rear is a threaded tension rod to give precise incremental adjustment for the tension on the drive cord to the bobbin and flyer. This example is stamped "E.L." on the top of the rear left leg. The bobbin and flyer units on Guilford chair wheels have two whorls on each so that there are four altogether. Since each whorl has a slightly different circumference, that gives four different combinations and allows for minor adjustments by the spinner. Magnificent finish. DW=14". *Courtesy of David Pennington.* $800-900.

67

What makes the Guilford wheels so special are not the simple square lines, but the tensioning systems and the fact that the bobbin and flyers have two whorls each. We noted earlier that most chair wheels lack a way of adjusting the tension on the drive belt or cord between the wheels. The Guilford type allows the front and back supports for the accelerating wheel to be raised or lowered. Also, the standard tension system for the drive cord to the bobbin and flyer employs a threaded nut that holds the "mother-of-all" in place once the spinner has pushed it to the appropriate tightness, a system that does not allow for incremental changes and is not very satisfactory. The Guilford crew found a way to adapt the threaded rod system used on saxonies so that incremental adjustments could be easily made. The square posts with only a few turned parts kept manufacturing costs down and give the Guilford pieces a wonderful, austere look. They were stained, sometimes brown, and in the case of the "G.W." example a dramatic cherry look was achieved (Fig. 6-16). One solid wheel example was painted red a long time ago, but not necessarily originally (Fig. 6-17).

Fig. 6-16. The original cherry stain on this Guilford example is breathtaking. It is so far the only one marked "G.W." on the top of the two front legs. The bobbin and flyer are old, but not original and have been converted to the typical Guilford four whorl system. The distaff pieces are replaced. DW=14". *Courtesy of Michael Taylor*. $700-800.

Fig. 6-17. Three of the reasons why Jim Munsie's chair wheel collection is considered exceptional. The one in old red paint is complete. Some wheels seem to find particular collectors. *Courtesy of James Munsie*. $900-1000 for the one in old red. The others have murky finishes and various repairs. $500-550.

Another group of chair wheels, which are found around the Lynn and Essex, Massachusetts area have a squared to round stock for the four legs (Fig. 6-18). These wheels have two different finial types, but we know that different finials were occasionally used by the same maker, so we think they were probably all from one source. These wheels have one technological feature of note. The hole for the front wheel post is elliptical and allows that post to slide from side to side, but the rear axle hole is round. This arrangement means that the accelerating wheel can twist slightly and twists the drive belt allowing for wheel alignment and tension adjustment. If these wheels do indeed come from the Lynn/Essex, Massachusetts area, can we still call the style Connecticut chair wheels? Why not? We call the Irish castle wheels of Lancaster County, Pennsylvania, Irish castle wheels. Another chair wheel type found frequently enough to warrant identification is that shown in Fig. 6-19. Somewhere in Connecticut or Massachusetts someone was producing these in quantities as we have seen five or six. The one shown in Fig. 6-20 can only be described as an "off-set" chair wheel as the accelerating wheel is noticeably off center. So far this example is unique. In looking at the chair wheels one is struck by the different places the distaff assembly is located. Some chair wheels have more than one hole for the distaff assembly.

Fig. 6-19. This style of chair wheel must have been made in relatively large quantities as five or six are known to us. It is very attractive. Notice that the distaff assembly is on the right side. Probably Connecticut in origin. DW=13". *Courtesy of James Munsie*. $750-800.

Fig. 6-18. This chair wheel was found in an attic in Lynn, Massachusetts. It must have been placed there soon after purchase as it has hardly been used. It closely resembles a chair wheel from the Essex Institute in Salem, Massachusetts. The accelerating wheel is actually larger than the drive wheel which is unusual. The distaff itself is a replacement. DW=12". *Courtesy of Michael Taylor*. $800-850.

Fig. 6-20. A wonderful example of the variety that the chair wheel form took. The off-set accelerating wheel is very distinctive. It is in excellent condition with a lovely finish. DW=13 ½". *Courtesy of James Munsie*. $800-850.

Fig. 6-26. These relatively tiny wheels are very transportable. Some are signed "J. Farnham" who holds an 1825 patent. Given the complexity of the treadling system, it was probably this type of wheel Joel Farnham patented rather than the more successful type shown in Fig. 4-10. Serious collectors really like these. DW = 8". *Courtesy of David Pennington.* $700-750.

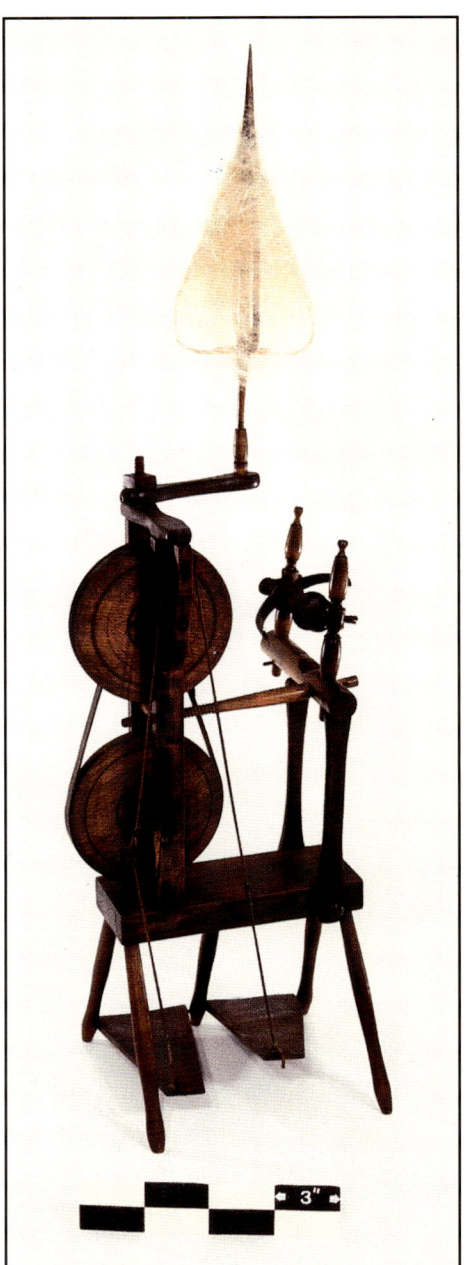

Fig. 6-27. A rear view of the previous wheel showing the odd arrangement for the treadles. *Courtesy of David Pennington.*

The wheel in Fig. 6-26 is not signed by Farnham but represents the type well. We know of more than 10 survivors of this form, so it must have been a relatively successful wheel in its day. Two of them are signed "J. Farnham." It is interesting to compare Figs. 6-26 and 4-12 and as one can see that in spite of the size difference, the tensioning systems are very similar. It is as if Farnham were using the double wheel system to shrink the size of the four-legged type to produce a transportable model. Anyone lucky enough to have one will have noticed how easy it is to pick up the five-legged model by its handle and carry it around. It would have been very easy to put in a carriage to go to a spinning bee. In this case the double treadle is helpful to maintain the power needed to keep the relatively light wheels going. The picture from the rear of this type (Fig. 6-27) reveals its real technological "advance," its way of handling the attachment of the two treadles to the axle. The total effect is quite comical. The footmen pull down alternately. On this type – we call them "little solid double's" – the right treadle attaches to a short footman at the bottom and the left treadle attaches to the top of the short footman by means of a long cord which goes over a pulley at the top of the wheel post in the back. This short footman is commonly missing. Instead the cord can be tied directly to the axle crank, and it works quite well.

Double treadles have other advantages too as the spinner can stop at anytime and then restart simply by pushing down on the appropriate treadle. With a single treadle the spinner often has to restart the wheel by hand. The example shown in Fig. 6-28 is a little mind-boggling as it appears to have once been a double flyer. It would take a heap of treadling to provide the torque to get it under way given how light and small the wheels are. While the holes are there on this wheel, it does not appear to have been used as a double flyer wheel. Perhaps, the right side of the superstructure was removed soon after purchase when it proved too difficult to use.

Several double flyer wheels employing a double treadle, double wheel arrangement have survived. One is based on the chair wheel (Fig. 6-21). With their solid wheels and overall chunkiness, they are not very attractive, but they are fascinating. They have a way to adjust the tension between the drive wheel and the accelerating wheel. At least, two of these have survived. Some of the Alpheus Webster type wheels are also double flyers, and the one in Fig. 6-29 is so close in turnings to the signed versions that it may well have come from Webster's shop.

Fig. 6-28. This example has a hole through the table on the right side and an extra threaded hole in the receiver between the wheels. Theoretically that would allow for a second structure on the right to house another bobbin and flyer unit. None of the other examples have these arrangements, so we don't know for sure if it was designed as a double flyer unit. DW=8". *Courtesy of Michael Taylor*. $700-750.

Fig. 6-29. This example is virtually identical to Fig. 6-7 except for the additional bobbin and flyer unit. It is probably from Alpheus Webster's shop. *Courtesy of David Pennington*. $800-850.

In central Pennsylvania, John Burley was making some of the most beautiful spinning wheels of the mid-19th century or was he? The wheel in Figs. 6-30 and 6-31 is certainly the most elegant American wheel we have ever seen. It is certainly signed "John Burley." So why question the attribution? That John Burley was involved is not the question, the question is who made the wheel. John Burley is listed in census records of Fayette City, Lycoming County, Pennsylvania as a blacksmith, and the metal work on this wheel is superb and even is signed "J. Burley." It is dated 1845 and numbered #13. Maybe someone else did the fine wood work, and Burley did the metal work and the marketing. The mark includes the word "warranted" (Fig. 6-32). But the maker of the works of a clock often had someone else make the case since the woodworking skills involved were different from the fine metal-working skills in clock-making. Also, we know that retailers of chairs had others make the chairs to which the retailer attached his name. John Burley "warranted" his wheels. Who did the woodwork on them is another question. That he is responsible for some of the best and most valuable American wheels is doubtless. Another question is why he indicated on his stamp what state he was from, but not what city. Is it possible that these were sold in New York City, Boston, Baltimore or Charleston, South Carolina. The wheel was certainly made for a high style setting. Perhaps they were for Philadelphia, and Burley was counting on some state pride to help sell his wheel.

Fig. 6-30. Oh my! What else can one say? Tiger-striped, beautifully made, a highly desired type, this wheel has always been the ultimate American wheel. It is John Burley's #13 wheel. His #2 once belonged to the Henry Ford Museum. These are the only two we know that have survived, and we have no idea what the total output was. This wheel was originally collected by the dealer/collector Bessie Spangler. Burley was a blacksmith working in Fayette City, Pennsylvania in the 1840s. This wheel has the date 1845 on it. DW=13". *Courtesy of Jeanne Asplundh.* $3500-4000.

Fig. 6-32. Close up of the "John Burley" mark. It is interesting that John Burley included the state in his mark which suggests that he may have been selling them outside of Pennsylvania. *Courtesy of Jeanne Asplundh.*

Fig. 6-31. Another angle on the previous wheel. It is amazing. *Courtesy of Jeanne Asplundh.*

Another version of this style of wheel was made by someone who carved his initials, "M.J.Y," into his wheels. Normally, we don't think of carved initials as maker's marks, rather they are likely to be an owner's mark, but we have seen three examples all similarly marked. Whoever "M.J.Y." was, he was also responsible for some of the very best wheels we have ever seen (Fig. 6-33). The one shown has number 383 on it and one numbered 381 with somewhat different finials turned up in South Pasadena, California. The one shown in Fig. 6-34 was unfortunately painted white before being stripped, but it is No. 335 and does have its original distaff assembly. These wheels have a Pennsylvania spoke and have been found in that area. They are beautifully made, especially the tensioning system for the belt between the two wheels, and they are excellent spinners. Another exceptionally nice example of this type of wheel (Fig. 6-35) is unmarked. Some rather clunky examples of this type of double treadle, double wheel flax wheel have also survived.

Fig. 6-33. This wheel marked with the carved initials "M.J.Y." is one of four such wheels, hence we believe "M.J.Y." is a maker's mark. This one is "No. 383". The distaff assembly is copied from No. 335 which is shown in Fig. 6-34. The finish on this example is excellent. D=14 ½". *Courtesy of David Pennington.* $1500-1700.

Fig. 6-34. Another marked "M.J.Y." wheel in complete condition but with a nasty stripped finish. A good example of how the current market favors finish over completeness. DW=14 ½". *Courtesy of David Pennington.* $800-900.

Fig. 6-35. Another unmarked version of the three preceding wheels, all of which almost certainly come from Pennsylvania. They are wonderful spinners. This one has a horizontal table supporting the mother-of-all. There are traces of a gilt paint which was probably the second finish. On top of that are traces of some much later red paint which has mostly been carefully removed. It is quite tall compared with the previous two examples and is more than eight inches taller to the top of the handle of the threaded screw which adjusts the tension between the wheels. DW=12 ¾" *Courtesy of Michael Taylor*. $850-950.

The wheel in Fig. 6-36 has a different configuration. It was found in southern Ohio where there was supposedly another one. Its wheel rims are so light that the wheel is very difficult to operate as there is little momentum to help the spinner in the treadling. It is no wonder that only one has survived.

The Canadians also made some interesting double treadle, double wheel flax wheels and adapted them to wool with large bobbin and flyers with big hooks and orifices. The example shown (Fig. 6-37 and 6-38) has a wonderful stamp, "J. L'Heureux." In some examples of his wheels the drive and accelerating wheels are the same size and in others noticeably different.[9]

These double treadle wheels with their accelerating wheels are interesting examples of the American fascination with ingenious improvements.

---

[1] Serge Daniloff, "Some Rare Spinning Wheels," reprinted in *The Art of the Weaver*, edited by Anita Schorsch, (New York: Universe Press, 1977): 21-24. It appeared in *The Magazine Antiques* in October, 1929.

[2] See the chapter on accelerating heads for a discussion of the success of Amos Miner's accelerating wheel head, which was made and sold in the hundreds of thousands during most of the 19th century.

[3] Florence Feldman-Wood, "Vertical Two-Wheel Spinning Wheels," *The Spinning Wheel Sleuth* 15 (January 1997): 8-10.

[4] To our knowledge the only other example of an A. Webster double treadle, vertical wheel is at the Farmer's Museum in Cooperstown, New York. It is misidentified as a Shaker wheel. Virginia D. Parslow, "Spinning Wheels," *Handweaver and Craftsman*, (Spring, 1956): 22.

[5] D. Demming, "Account of the New York Accelerating Wheel-Head," *Archives of Useful Knowledge*, Vol. II, No. 2 (October, 1811): 106.

[6] Florence Feldman-Wood, "Vertical Two-Wheel Spinning Wheels": 9.

[7] Florence Feldman-Wood, "Guilford-Style Chair Wheels," *The Spinning Wheel Sleuth* 26 (October, 1999), 2-4.

[8] Eliza Leadbetter, *Spinning and Spinning Wheels*, Shire Album 43, Shire Press, 1979: 19.

[9] Judith Buxton-Keenlyside states that L'Heureux was from L'Acadie, Quebec. On the example she shows the accelerating wheel (5¾") less than one half the size of the drive wheel (15½") whereas on the one see here, the accelerating wheel (14½") and the drive wheel (14½") are the same size. She shows a closely related wheel marked "M.R..". Judith Buxton-Keenlyside, *Selected Canadian Spinning Wheels in Perspective: An Analytical Approach* (Ottawa: National Museums of Canada, 1980): 177-180.

Fig. 6-37. This French-Canadian wheel from near Montreal has the date 1888 penciled on the bottom of one of its treadles. It is boldly marked "J. L'Heureux". It has a big orifice, large hooks, and no distaff hole, so it was probably made for spinning wool. There are some versions of this wheel marked "M.R.", some with the drive wheel underneath the table. DW=14 ½" *Courtesy of Jeanne Asplundh.* $600-650.

Fig. 6-36. This small double treadle, double wheel example has a different configuration than any of the preceding. The wood used in the thin rims is very light, and the wheel is very hard to use as there is little momentum. The wheel was found in Southern Ohio, and there was supposed to be another one "just like it" there. DW=15". *Courtesy of David Pennington.* $500-550.

Fig. 6-38. A close up of the "J. L'Heureux" mark from the previous wheel. What appears to be a tensions crew is simply a handle for carrying the wheel. The metal band around the mother-of-all canbe loosened or tightened, allowing the mother-of-all to be tilted for tensioning purposes. This type of tensioning is also found on many Canadian saxonies. *Courtesy of Jeanne Asplundh.*

Chapter 7
Double-Flyer Wheels

The first evidence of a double-flyer wheel is in Thomas Firmin's pamphlet, "Some proposals for the employment of the Poor and for the prevention of Idleness and the consequences thereof, Begging." published in 1681, in London.[1] Firmin both describes and illustrates such a machine. Some have speculated that it was the introduction of the fly-shuttle loom in the 1730s that spurred the use of such wheels, but there is scant evidence of their use in the American colonies. Robert G. Stone looked at a sample of 978 probate inventories containing textile tools from 1720-1819 in Woodbury, Connecticut. The first double wheel appears by name in 1775. From 1775 to 1794 only four such wheels are noted. From 1795-1819 forty-nine such wheels are listed.[2] Baines in her book, *Spinning Wheels, Spinners and Spinning*, mentions that double flyer wheels were not mentioned much until the late 18th century in Europe either.[3] There are enough different varieties of double-flyer wheels that a collector could have a good-sized collection of just them.

Fig. 7-1 shows an "S. Barnum" double-flyer wheel. Similar wheels are signed "J. Sturdevant." Silas Barnum (1775-1828) and John Sturdevant, Jr. (1760-1825) married sisters (?). John Sturdevant, Jr. had a son, John Sanford Sturdevant, and Sanford is the family name of a prominent Connecticut family of double-flyer wheel makers. Given their proximity to one another in the late 18th century, they are likely to have influenced one another.

These wheels are vertical with the wheel above a bottom table and a frame of four posts supporting a top table. The bobbin and flyer units are above the top table, suspended from front maidens and a back crossbar, which can be raised or lowered by two threaded rods to change the tension. The lower table has four legs and a single treadle. There are short vertical wheel posts front and back. This particular arrangement is somewhat reminiscent of Firmin's 1681 drawing, which is, however, without a top table. In addition to extensive decorative black score marks, Barnum and Sturdevant (and

Fig. 7-1. A very nice example signed "S. Barnum" of one of the more common types of double flyer wheel. In America double flyer wheels were almost all vertical with most having the wheel below, but some have the drive wheel above the bobbin and flyer units. On a double flyer wheel the bobbin and flyers should be identical, and one usually has an "X" carved on the flyer. Double flyers without the original bobbin and flyer units are very expensive to restore and are not worth more than $200-250. DW = 18 ½". *Courtesy of James Munsie.* $650-700.

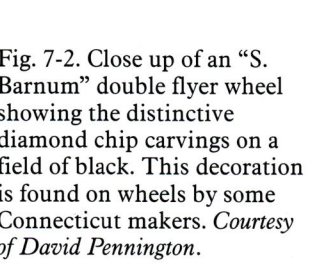

Fig. 7-2. Close up of an "S. Barnum" double flyer wheel showing the distinctive diamond chip carvings on a field of black. This decoration is found on wheels by some Connecticut makers. *Courtesy of David Pennington.*

some other Connecticut makers) used chip carvings in the shape of diamonds on fields of black (Fig. 7-2). Most double-flyer makers marked one of the pair of bobbin and flyers with an "X" on the flyer, but we aren't sure why.

Solomon Plant (1741-1822) was an early maker of double flyer wheels, who left his account book, which is in the collection of the Stratford Historical Society, Connecticut. "S. P." marked double-flyer wheels exist and between 1810 and 1821 Solomon Plant's account book shows that he made 160 "dubles," 56 singles and 55 great wheels. It appears that his doubles were also worth $4 although he continued to list prices in pounds and shillings (1:4:0 for a "duble"). Two other interesting things can be gleaned from his account books. He distinguished between "duble rims" and "single rims" in his book. The Stratford Historical Society has both a double-flyer wheel and a saxony marked "S.P.," and the saxony drive wheel is 2 inches larger than the double-flyer drive wheel. He sold 81 of the 160 doubles he made between 1810 and 1821 to two merchants in Fairfield, Connecticut.

The Sanford clan of Newtown, Connecticut made a similar double-flyer wheel. At least four members from three generations of Sanfords made such wheels, and one Elias Bristol Sanford went his own way and patented a distinctively different style of double-flyer wheel. Samuel Sanford (1743-1817) of Newtown, Connecticut at death had ten "double wheels, $40" in his inventory. (From the records of other double-flyer makers at this time, it appears that $4 was the going rate.) Samuel had a son, Isaac (1768-1748) by his first wife, and another son, Josiah (1793-1851), by a second wife. Josiah is said to have succeeded his father in his spinning wheel business. A cousin, Beardsley Sanford, made double flyer-wheels, marked "B. Sanford," and this mark should not be confused with "E.B. Sanford." Beardsley moved to Fergusonville, New York and made a variety of wheels there, which we discuss in Chapter 10. Reports of a "D. Sanford" mark are almost certainly based on a broken "B. Sanford" stamp. The Sanford double-flyer wheels are consistent in having their drive wheels supported by cross-stretchers between the front posts rather than having short vertical posts. I. Sanford wheels have two distinctly different maidens and two different types of wheel spokes.

Wheels identical to the unmarked one shown in Fig. 7-3 are marked "I. Sanford" and "S.S." (Samuel Sanford we assume).[5] A double-flyer marked "B. Sanford" has the same maiden, but a plain spoke. Some "I. Sanford's" (Fig. 7-4) have the plain spoke and an urn-shaped maiden finial, which is what most "J. Sanford" (Fig. 7-5) double-flyer wheels look like. One "J. Sanford" has a very plain lozenge finial.

Fig. 7-3. An unsigned example of the Sanford clan's work. It is included because an identical wheel marked "S.S." is shown in Leadbeater's book and Samuel Sanford was the clan patriarch. Another identical example marked "I. Sanford" is shown in Florence Feldman-Wood's classic article on double flyer wheels. Isaac was Samuel's oldest son in the business. Later Isaac Sanford produced doubles with different finials and spokes. DW=17 ½". *Courtesy of Michael Taylor.* $500-550.

Fig. 7-4. Marked "I. Sanford" this double flyer has plain spokes and urn finials. Other examples with these features have been spotted. The likely explanation is that Isaac was changing as tastes changed. His much younger half-brother, Josiah, made wheels like this one. DW=18". *Courtesy of David Pennington.* $500-550.

Fig. 7-5. This double flyer marked "J. Sanford" is in the best condition imaginable. Possibly never used. The wheel in Fig. 7-3 shows extreme treadle wear whereas this one shows none. The hooks on the flyer arms show no wear. DW=17". *Courtesy of David Pennington.* $600-700.

All of these Sanford wheels look quite similar to most non-collectors because of the horizontal wheel supports. The changes in turnings reflect the change of style in furnishings in this period with elaborate turnings giving way to very plain ones between 1810 and 1830. One marked "Russell Field" (Fig. 7-6) and one marked "L. Judson" (Fig. 7-7) also have horizontal wheel supports.

The most dramatic of the Connecticut double flyer wheels with the drive wheel below is a type which we used to call "log tops." The one marked "J. Miles" (Fig. 7-8) is a spectacular example of this small group. (The J. Miles stamp is also found on some very plain chair wheels Fig. 6-12.) Another nice example is marked "D. Bird." Bill Ralph makes the point that the additional threaded screws allow for the separate adjustment of the front maidens, which is an advantage over the more traditional set up favored by other makers.[6] Some people must have wanted this feature as several unsigned examples as well as these two signed examples have survived.

Fig. 7-6. This wheel is stamped "Russell Field" and resembles the work of the Sanford family. It is a very nice example. DW=17". *Courtesy of David Pennington.* $500-550.

Fig. 7-7. This wheel is stamped. "L. Judson" and is also similar to the Sanford clan's double flyer wheels. It does not have any of the distaff assembly. DW=17 ½". *Courtesy of David Pennington.* $450-500.

Fig. 7-8 shows what we call a "log top". It is a desirable variation of the American double flyer wheel with the drive wheel below the bobbin and flyers. It is marked "J. Miles" and is complete with a fine old finish. DW=18". *Courtesy of David Pennington.* $850-900.

Fig. 7-9. "E. B. Sandford" took out a patent in 1816 and this wheel is signed "E.B. Sanford". Five examples are signed and two are not. Most, like this one, have twelve spokes, but one signed example has fourteen (Fig. 7-11). *Courtesy of Jeanne Asplundh.* It is in excellent condition. DW=17". $1100-1200.

Fig. 7-10. This close up of the wheel in Fig. 7-9 shows the "E. B. Sanford" stamp and the many oddities on this patent wheel. It has a tensioning alignment screw from front to back. It has metal flyers with hooks. The bobbin has two whorls and the flyer only one whorl. The black marks on the wheel post are decorations, not staining from use. *Courtesy of Jeanne Asplundh.*

Elias Bristol Sanford (1791-1881) was the son of Isaac Sanford and would have been a contemporary of his uncle, Josiah. A spinning wheel patent was granted to an "E.B. Sandford (sic)" of Newtown, Connecticut in 1816, which is almost certainly Elias Bristol Sanford in spite of the extra "d" in the patent records. Elias Bristol was in Newtown, Connecticut until at least 1819 when his first wife died there. Several unusual double-flyer wheels with the mark "E.B. Sanford" have turned up (Fig. 7-9). Two or three unmarked examples also are known. These wheels have several unusual features; but since the patent records burned in 1836, we do not know any details of what was patented. The most interesting feature is the little threaded rod which goes from the front post to the back post. The side view in Fig. 7-10 will help with this feature and others mentioned. Because the hole drilled in the back post is slightly off center, tightening this screw causes the front and back post to shift in alignment with each other and not to simply come closer together. This device allow for aligning the drive cord. It may also give some fine tuning of the tension. The metal flyers and the way the hooks are attached are unusual on an American wheel. The way the bobbin and flyer units are held in place in the back by "thumbscrews" with leather liners is unique. Another notable feature is that there are two bobbin whorls and only one flyer whorl. The tension device is odd and not very satisfactory. The spinner simply loosens the threaded nut in front and pushes down until the desired tension is achieved and then tightens the nut. The reader, by now, is probably wondering why we haven't mentioned the most obvious difference between E.B. Sanford's double-flyer wheel and those of his relatives, which is that the drive wheel is above the bobbin and flyer. There are a number of double-flyer wheels with the wheel above the bobbin and flyers, which we will note later in this chapter. E.B. Sanford also decorated his wheels with chevrons of graining on the square side of the front wheel support post. This graining can be mistaken for the dark marks caused by linseed oil stain thrown off from flax hooks unless you get a virtually unused wheel (Fig. 7-10) in which case the chevron effect is quite noticeable.

One "E.B. Sanford" wheel (Fig. 7-11) has a drive wheel with fourteen spokes whereas the rest all have twelve spokes. That raised a question about how these wheels were assembled. Did E.B. Sanford have more than one source for his drive wheels? Is the fourteen spoke model an earlier one, and he later decided he only need twelve? Definitely an odd development.

Earlier we mentioned that unmarked wheels identical to those of "S.S." and "I. Sanford" are known. Some Shaker wheels are marked and some are not. Some clones of "A. Webster" double treadle, double wheels are known. Why would someone with a stamp for marking wheels not use it? Our working hypothesis is that spinning wheel makers marked the wheels they left with retailers as a kind of informal warranty system or perhaps as a kind of advertising/branding. Those who bought directly from the maker would not need the warranty since they had purchased directly from the maker.

Fig. 7-12. An "attic found" wheel in pristine condition except for the missing distaff, this double flyer wheel still had old flax on its bobbins. Many examples of this particular wheel form have been collected over the years, but only one was marked ("T. A."). There is only one tension screw, so we can be confident that a single, doubled drive cord was used. DW= 18". *Courtesy of Michael Taylor.* $750-800.

Fig. 7-11. This signed example of an "E. B. Sanford" double has fourteen spokes instead of the usual twelve which raises questions about whether Elias Bristol Sanford was having others make his patent wheels while he was marketing them as well as making them. DW=17". *Courtesy of Jeanne Asplundh.* $950-1000.

The most common kind of double-flyer wheel (Fig. 7-12) with the wheel above is almost never marked (although one example marked "T. A." is known). More than a dozen of these have been spotted over the years and their distinctive design is worth noting. The long threaded rod coming up from the low table which moves the bobbin and flyer assembly up and down is a feature as are the flame-shaped finials on the four maidens. Some double-flyer wheels were set up to have two separate drive cords. If so, there are independent tensioning screws for each bobbin. Usually one bobbin and flyer unit will be noticeably offset from the other so that the two separate drive cords do not become tangled. In Fig. 7-12 the double-flyer wheel has only a single tensioning device, so it would have one (doubled) drive cord. The next double-flyer wheel we look at clearly had two separate drive cords.

The wheel in Fig. 7-13 is quite unusual. Only two have been seen. One is in the Cummer collection at the American Textile History Museum. The other is in Mike's collection and was found in eastern Rhode Island. The separate tension devices are evident. When it is looked at from the side, it is easy to see that one of the bobbin and flyers is farther forward than the other. The drive wheel was originally painted red, probably to discourage splitting and checking as the wood dried, but possibly for decoration. Sectioned drive wheels are almost never painted, but if the maker knew the wood was a little green, he might have taken a little extra precaution.

Fig. 7-13. Every collector has "favorites" and this one is Mike's. One of only two known examples of this style. It has independent tension screws and the bobbin and flyer units are offset so that two independent drive cords can be used. As Joan Cummer pointed out, double flyer wheels with independent tensioning systems may well have used two separate drive cords. DW=21 ½". *Courtesy of Michael Taylor*. $800-900.

Fig. 7-14. This example with a small table for the maidens is very unusual. It is in beautiful condition! The turnings, as well as the style, scream "Connecticut". Among the very best spinning wheels in every respect. DW=18" *Courtesy of David* Pennington. $850-950.

In our first book we suggested that the wheel in Fig. 7-14 was not uncommon as far as double-flyer wheels go. Boy, were we wrong. It is a very much like the first "T. A." double, but it has a second table. It is distinctly different from the "Andrew Beers" examples.[6] The finials are similar to the early Sanford doubles, but the spoke is simple. It is likely from western Connecticut around 1800-1820 and is a wonderful wheel in super condition.

In Europe the small upright "parlor" wheels and the upright Swiss wheels also became double-flyers (Figs. 7-20 and 7-21). Even saxonies were made with double-flyers both in the British Isles and on the Continent (Figs. 7-22 and 7-23). There is a famous picture of Queen Victoria seated in front of one although as an "amateur" spinner it is not surprising that she has had one of the bobbin and flyer units removed.[7]

---

[1] Patricia Baines, *Spinning Wheels, Spinners and Spinning*, (New York: Charles Scribner's Sons, 1977), 149-151. A close contemporary of Firmin, Randle Holme, also mentions "double spool" wheels by which two threads of worsted ("Jersy") could be spun at once. Holme published his work in 1688 but began work on it in 1649. Alan Raistrick, "Extracts from the *Academy of Armory* by Randle Holme pf Chester, 1688, *The Spinning Wheel Sleuth*, 20 (April 1998), 2-4. It is not clear if the "double spool" wheel of Randle's is vertical like Firmin's illustration, or is like a saxony with a sloping table.

[2] Florence Feldman-Wood, "Double-flyer Spinning Wheels," *The Chronicle of the Early American Industries Association*, Vol. 53, No. 4 (December 2000), 132-133.

[3] Patricia Baines, *Spinning Wheels, Spinner and Spinning*, 151.

[4] Florence presents much of the same material on Connecticut double-flyer wheels from the aforementioned article in *The Chronicle of the Early American Industries Association* in two articles in *The Spinning Wheel Sleuth*. Florence Feldman-Wood, "Sanford Family Spinning Wheels," *The Spinning Wheel Sleuth*, 30 (October 2000), 5-6 and Florence Feldman-Wood, "Solomon Plant, Wheel Maker of Stratford, CT," *The Spinning Wheel Sleuth*, 31 (January 2001), 2-4. All are worth careful study by those interested in double-flyer wheels.

[5] The "S.S." double-flyer wheel is shown in Eliza Leadbeater's *Spinning and Spinning Wheels*, 18, and appears to be identical to marked I. Sanford one shown in Feldman-Wood's article "Double-flyer Spinning Wheels," 134.

[6] Bill Ralph, "An Unusual Dual Spindle Wheel," *The Spinning Wheel Sleuth* 19 (January 1998): 5-6.

[7] Michael Holcomb, "Signatures" *The Spinning Wheel Sleuth*, 20 (April 1998): 10-11.

[8] Patricia Baines, 174.

Fig. 7-20. A nicely painted example of a European parlor wheel with two bobbin and flyer units. *Courtesy of Joe Roby*. $300-350.

Fig. 7-22. Double flyer saxonies from Great Britain and Holland occasionally surface, but so far there are no examples from America. Queen Victoria was photographed using one with only one bobbin and flyer in place. The example shown here is in exceptional condition. DW=17". *Courtesy of Jeanne Asplundh.* $900-1000.

Fig. 7-21. A side shot of the European double flyer in Fig. 7-20 showing one very unusual feature, the flyer has hooks on all four sides! From this perspective the viewer can see that the bobbin and flyers each have their own drive cords. *Courtesy of Joe Roby.*

Fig. 7-23. Another view of the wheel in Fig. 7-22. The drive wheel on such doubles must be very wide to accommodate two separate drive cords, and the bobbin and flyer units have to be offset from one another as this picture shows. *Courtesy of Jeanne Asplundh.*

Chapter 8

Accelerating Wheel-Heads "The Pleasant Spinner," Amos Miner, and the "Minor's head"

"The Pleasant Spinner" was the name that Daniel Read, of Brookfield, New York, gave to his patent spinning wheel that he applied for on September 10, 1811.[1] We are aware of the details of this patent because of a copy of it turned up as part of an unsigned, undated letter of conveyance in files at Canterbury Shaker Village in Canterbury, New Hampshire. The patent description was provided by Florence Feldman-Wood, editor of *The Spinning Wheel Sleuth*. The appearance of this patent solved several mysteries.

In our previous book, *A Pictorial Guide to American Spinning Wheels,* we showed a version of this wheel in our section on patent wheels on a hunch (Fig. 8-1).[2] We knew they were found at the Pleasant Hill Shaker community in Kentucky, but we also knew that this type of wheel had been found in various forms in other places as well (Fig. 8-2). Since we also knew that the patent office had burned in 1836 and that most wheels patented before that time were known only from a list which included "invention, inventor, residence, date and no.," we assumed that this type of wheel was probably in that group.[3] On visits to various Shaker sites we were often asked what we knew about "pleasant spinners," which were occasionally mentioned in Shaker records.[4] To which we could only say, "nothing." Patent records normally contain both a description and a drawing. In this case only the description was available.

Fig. 8-1. Ultimately we found out that this wheel is one of the many forms taken by "wheel-less wheels" or "Pleasant Spinners". In 1811 Daniel Read of Brookfield, New York took out a patent for the "Pleasant Spinner", a clever adaptation of Amos Miner's patent. These wheels are powered by some other power source as the large whorl or whorls on the axle of the wheel show. *Courtesy of David Pennington.* $150-250.

Fig. 8-2. Seeing this "Pleasant Spinner" with a base was a big "aha" experience as it indicated what type of base might have gone with a wheel like Fig. 8-1. Five of us have been collecting for over 30 years each, and this is the only one with a base that we have seen. Perhaps they were sold without bases and it was up to the buyer to supply one. W=13". *Courtesy of Jeanne Asplundh.* $350-400.

Fig. 8-3. Another example of a Pleasant Spinner from the Elkins, West Virginia, area where a collector of spinning wheels in the 1960's and 70's claimed to have seen more than a dozen of this type. W=13 ½". *Courtesy of Michael Taylor.* $200-250.

Mike realized that the patent record described, almost exactly, a small "wheel-less wheel" he kept at his office. (Fig. 8-3) The wheel diameters differed slightly (15" in the patent, about 10]!" in Mike's), but Read's patent clearly stated that the wheel could be "of a larger or smaller radius." A very similar one with the same size of wheel is part of a group found in Indiana while Mike's example came from the Elkins, West Virginia area (Fig. 8-4). In Fig. 8-5 we show another variation.

Fig. 8-4. Collected out of Beverly, West Virginia, next door to Elkins, West Virginia along with the previous example. Different enough to suggest a different maker, but near enough by to suggest that this was a popular style in this region. W=14". *Courtesy of Jeanne Asplundh.* $200-250.

Fig. 8-5. Another variation on the wheels shown in Figs. 8-3 and 8-4. These wheels can be powered by a long thick cord attached to a fixed pulley. The cord was pulled or walked by the spinner to provide the power normally given by a drive wheel. Examples are found throughout Ohio, Michigan, and Indiana as well as associated with Shaker settlements from Maine to Kentucky. Daniel Read's patented "Pleasant Spinner" was quite popular. W=14". *Courtesy of Jeanne Asplundh.* $200-250.

Fig. 8-6. Another "wheel-less wheel" mounted on a base. The base looks far cruder than the wheel. Fastening the wheels to something is critical and a regular user might want a regular base rather than the dining room table. This example has a second accelerating wheel. W=17" *Courtesy of David Pennington.* $350-400.

Fig. 8-7. Clunky bases represent an opportunity for the skilled turner and here is an example of what at a first glance looks like a small wool wheel but is, in fact, a "wheel-less wheel" on legs. The large whorls next to the wheel give it away. W=22". *Courtesy of David Pennington.* $350-400.

Fig. 8-8. This style of "wheel-less wheel" was stuck in the floor and used as shown. There is a nail at the tip of the narrow end the keep it from sliding on the floor, but exactly how it was used is still not clear. W=13". *Courtesy of Jeanne Asplundh.* $200-250.

Read's "Pleasant Spinner" looks, at first, as if it could be a little table wheel which could be turned by hand. They do not, however, have hand cranks. It is the presence of the extra whorls on the drive wheel hub that is the give away. There are a number of varieties of these wheels, which as a class we call "wheel-less wheels" because on them, unlike the Miner's heads, there is no drive wheel intended (Figs. 8-6 and 8-7). Some other dramatic variations are shown in Figs. 8-8 and 8-9.

In Read's patent there is a description of how the wheel is to work. The spinner attaches a pulley to a convenient surface, runs a loop of rope from that pulley to the spinning apparatus and back, and can power the accelerating wheel by pulling on the rope or by simply holding the rope with one hand and walking while spinning with the free hand. We have shown several examples of "wheel-less wheels" as well as accelerating heads. Some of them are attached to their bases, and one looks very much like a small wool wheel.

Read's "Pleasant Spinner" is an ingenious adaptation of the principle of the accelerating wheel, which was patented by Amos Miner in 1810. It appears that Miner had already patented a spinning wheel with an accelerating head in 1803, but in 1810 he patented it in a form that allowed the buyer to keep the old wool wheel and simply insert the new accelerated wheel head into the hole in place of the old head. Interestingly enough, his partner, Davis Demming, states that Miner received his first patent in November, 1805 whereas the patent list shows it as November 16, 1803. Since Demming is quite explicit about the three years needed to make the necessary improvements, we suspect that the patent list is incorrect on the date and that Miner's first patent was in 1805.[5] Many wool wheels already had removable heads.

Fig. 8-9. This Pleasant Spinner variation is vertical like the previous one, but it was probably intended to be used on a table as it has a clamp. W=14". *Courtesy of Michael Taylor.* $150-175.

We occasionally find unaccelerated bat's heads (Fig. 8-10) and other forms of unaccelerated heads (Fig. 2-3) on old wheels or in piles of odds and ends in antique stores (Fig. 8-11). Miner's heads were wildly successful for the next 60 years and hundreds of thousands were made (Fig. 8-12).[6]

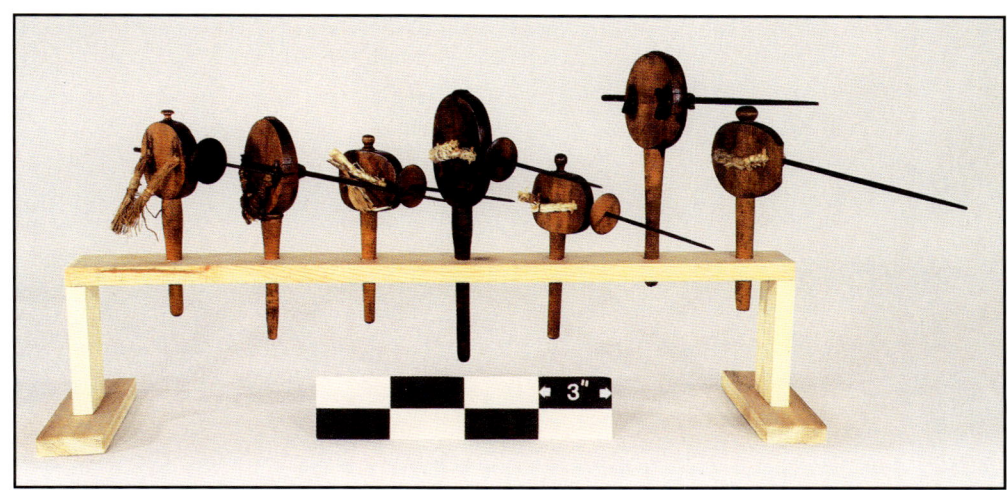

Fig. 8-10. A nice row of bat's heads. They are a wonderful find in an antique shop as wool wheels are often missing their "heads". They seem to be more common in New England than in Pennsylvania, but they turn up in both regions regularly. *Courtesy of David Pennington.* $60-70 with spindle.

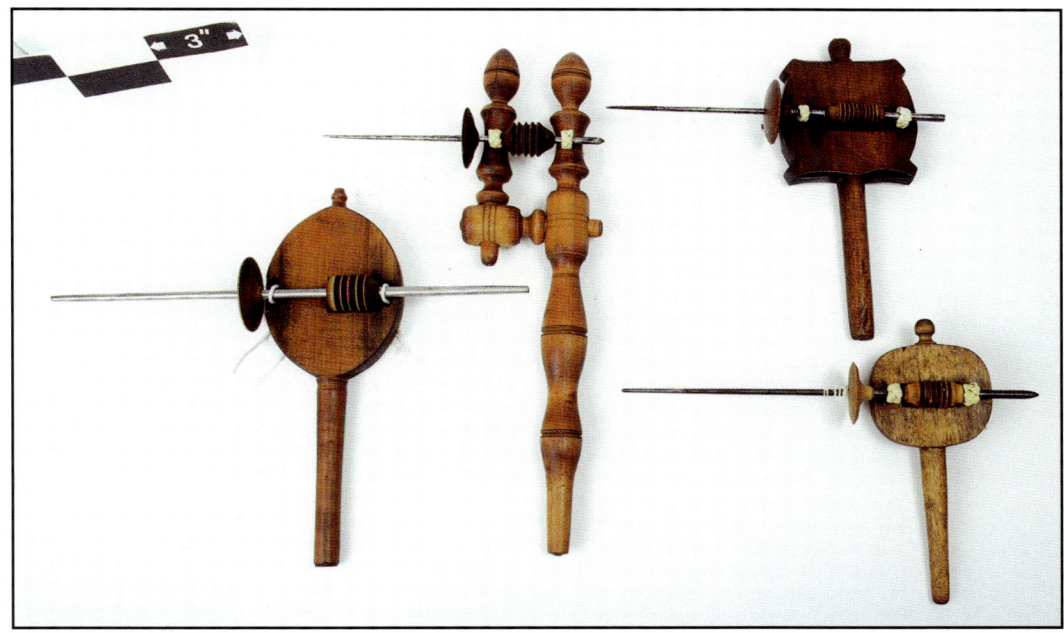

Fig. 8-11. Another nice assemblage of removable direct drives and bat's heads. The bat's head on the far right is from the Canterbury, New Hampshire Shaker Community. *Courtesy of Jeanne Asplundh.* $70-80.

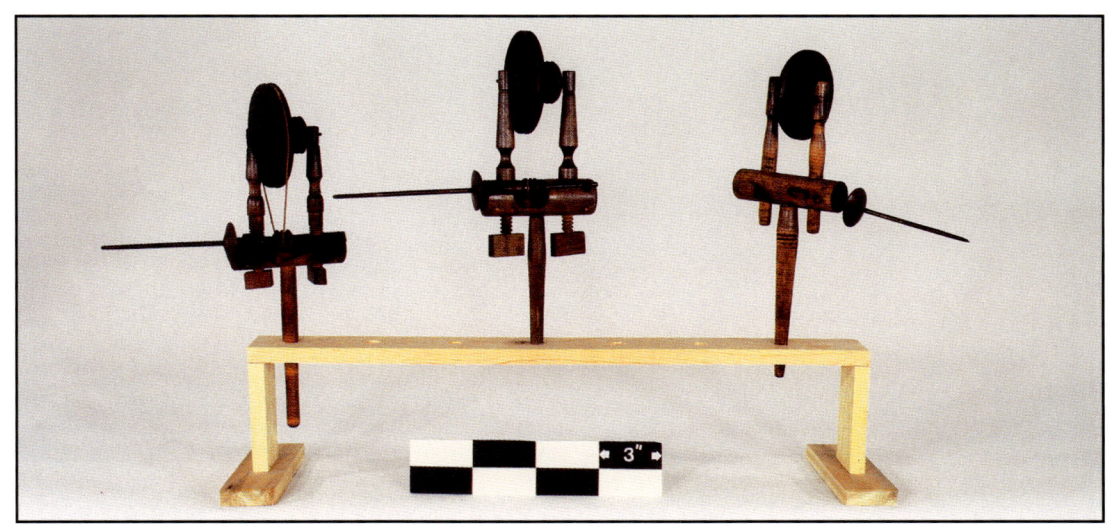

Fig. 8-12. Three Minor's heads made by different companies, mainly in the Chesterfield, New Hampshire area, over the course of the 19th century when literally tens of thousands of these were made. They are virtually identical to the drawing made by Miner's partner Demming in his 1809 article. They pop up on e-Bay regularly, missing various pieces. *Courtesy of David Pennington.* A complete one with a label intact indicating the maker and having directions for use, $95-100. Complete without a label but including the accelerating wheel and the spindle, $80-90.

Somewhere along the line, one of the holders of the patent rights (Miner sold out early on) misspelled Miner's name as Minor on the labels applied to these heads. His name is spelled both ways in the patent list, Minor in 1803 and Miner in 1810. So one is forever explaining to people that their "Minor's head" is really a Miner's head in spite of what it says on the old labels (Fig. 8-13).[7] The multiplying effect of the accelerating head on a Minor's head improved the spinner's efficiency 9-fold. The heads in Fig. 8-14 show how well the term "Yankee ingenuity" applies to our New England forbearers as they sought to improve on the Minor's head form; but the standard Minor's head, almost unchanged from the etching in the October, 1811 issue of *The Archives of Useful Knowledge*, is by far the most common form of accelerated head regardless of whom made it (Fig. 8-12).[8] He had produced an exceptional improvement.

Fig. 8-13. Close up of a Minor's head label. These paper labels are frequently missing. A Minor's head with an intact label and the spindle is a good find. *Courtesy of Jeanne Asplundh.* $95-100.

Right:
Fig. 8-14. Yankee ingenuity at work. Improved Minor's heads of various sorts. *Courtesy of David Pennington.* $100-110.

Chapter 6, on double treadle, double wheel flax wheels, is really an extended treatment of how Miner's accelerating head principle was applied to flax wheels. The obvious application of the principle is shown in Figs. 8-15 and 8-16. It is a saxony with an accelerating head. We have seen only two of these in over 30 years of collecting. Given the popularity of the accelerating wheel on both wool and flax wheels, the scarcity of these is a major puzzle.

Fig. 8-15. An extraordinary saxony wheel with an accelerating head. It is marked "H.M." and probably from just north of Boston. The obvious way to apply the accelerating wheel to a flax wheel is shown here. So far it is the only example we have seen. Since most accelerated flax wheels have two treadles, treadling was likely an issue. *Courtesy of Patricia A. Smith.*

Fig. 8-16. A close up of this wonderful wheel. The tension between the accelerating wheel and the drive wheel is adjustable. *Courtesy of Patricia A. Smith.*

The accelerating head was also adopted abroad. While we cannot prove that Amos Miner was the first to use an accelerating head on a spinning wheel, there are no earlier documented earlier foreign examples. The one shown in Fig. 8-17 is a Turkish accelerated flax wheel, which are now being imported widely. Some have distaffs and some do not. Most have a single cord going from the accelerating wheel to the flyer but with no bobbin whorl. It has been suggested that when the spinner holds the yarn that retards the bobbin enough for the wheel to function as if there were a bobbin brake.

---

[1] Michael Taylor, "The Pleasant Spinner," *The Spinning Wheel Sleuth* 27 (January 2000): 6-8.

[2] David Pennington and Michael Taylor, *A Pictorial Guide to American Spinning Wheels*, (Sabbathday Lake, Maine: The Shaker Press, 1975): 78.

[3] M.D. Leggett compiled a list in 1874 based on a variety of existing records including some which preceded the Patent Office fire of 1836. M.D. Leggett, compiler, *Subject-Matter Index of Patent for Inventions Issued by the U.S. Patent Office from 1790 to 1873. Inclusive.* Washington: Government Printing Office, 1874.

[4] Beverly Gordon, *Shaker Textile Arts*, (Hanover, New Hampshire: The University Press of New England, 1980): 49.

[5] D. Demming, "Account of the New York Accelerating Wheel-Head for Spinning Wool," *Archives of Useful Knowledge*, (October, 1811): 105.

[6] Frank White, "Heads Were Spinning: The Significance of the Patent Accelerating Spinning Wheel Head," *Annual Proceedings of the Dublin Seminar*, (Boston, 1999): 64.

[7] The solid wheel is usually painted black. These wheels were turned "green" on a lathe. They were painted to keep them from splitting as they dried, but many of these wheels have split and pulled part to the point where it appears that a large piece of the wheel has chipped out. On some mother-of-all's on saxonies, which were almost never painted, one can see the same big cracks, especially on the saxonies of Samuel Humes of Lancaster. The New Lebanon Shaker wool wheel hubs were painted red to prevent checking. Alpheus Webster, who used a similar solid accelerating wheel on his double treadle, double flax wheels also painted his black. The accelerating wheels on the Minor's heads made by Joel Farnham and labeled "J. Farnham" are sometimes painted green.

[8] D. Demming,104.

Fig. 8-17. This Turkish 'flax' wheel is hand-cranked with an accelerating wheel. DW=9 ½". *Courtesy of Jeanne Asplundh.* $350-400.

Chapter 9
Patent Wheels

Existing wheels for which there are patents, either in the United States or Canada, are a select group. There are many more patents in the respective patent files, but they may not have ever been made. Here patents of which there are no known examples are discussed.

The most important patent wheel is the 1805 mystery wheel of Amos Miner. One of his partners, Davis Demming, wrote a letter to the *The Archives of Useful Knowledge* in 1810, about six months after Miner received a second patent for his famous accelerating head, in which he states that "Mr. Miner, the patentee and inventor, in November 1805, obtained a patent for a double-geered (sic) great wheel, also for a horizontal little wheel: the accelerating wheel was in that model in an intermediate situation."[1] This wheel became the basis for a whole host of copies, but we know nothing more about what Minor's wheel looked like because in 1836 the Patent Office burned and destroyed the records. Perhaps it looked like Fig. 8-16. Inventors were invited to resubmit, but most did not. So what little we know about pre-1836 spinning wheel patents is limited to an "index" published after 1873 which listed the "invention, inventor, residence, and no.," and a few fortuitous discoveries.[2] Amos Miner is listed twice, once as Miner and once as Minor. It is certain that in April, 1810 he received a patent for his famous accelerating wheel-head, and thanks to an engraving of the accelerating head published with Davis's letter, we know what it looked like.[3]

Even the patent date of Miner's first wheel is disputed since the patent list states it was in November, 1803 while his partner says it was in November, 1805. It would be easy to misread a handwritten 1805 into an 1803. Demming speaks of the three years Miner needed to improve his first patent before engaging in the manufacture of it in 1809, which makes the 1805 date more likely. Since the name is also misspelled as Minor, it is possible that the editor of the index was working with a poor text.

The earliest patent wheels about whose appearance we have some information are those of Alpheus Webster and W.H. Peabody. The story of Alpheus Webster we take up separately in other chapters. He patented wheels in 1810 and 1812. We have seen two very unusual types of wheels with his stamp on them (Figs. 6-2 and 6-3), so we think we know what he patented but can't be sure. A drawing of W.H. Peabody's wheel is in the patent records. A few items had been taken out of the Patent Office the night of the fire by a clerk. There are no written specifications available for the Peabody wheel, but it looks very much like what Demming describes as the general configuration of Miner's patent wheel. Someday someone may find a copy of Minor's original patent for a spinning wheel in some old files as has happened with Read's "Pleasant Spinner" and Elijah Skinner's wheel.[4]

What makes Miner's original contribution so intriguing is that prior to it there is no evidence of the use of an accelerating wheel with domestic spinning wheels. Soon after Miner's accelerating head appears, spinning wheels incorporating that principle begin appearing on both sides of the Atlantic.[5] Peabody's 1812 wheel exists only as a patent drawing. Elijah Skinner's 1818 patent wheel was actually made and a few examples have surfaced. Craig Evans, curator of the Sandwich (New Hampshire) Historical Society, discovered a copy of the written description part of the patent.[6] A drawing of Skinner's wheel was one of the few items to escape the 1836 fire, so at last the two were united. The only complete example is at the Whaling Museum in New Bedford, Massachusetts. It has a spindle mechanism whereas the patent drawing shows a bobbin and flyer mechanism. The one at the American Textile History Museum was missing its spinning mechanism entirely when it was collected.

Evans' research turned up additional information about inventor's in the 19th century. Skinner did not manufacture his inventions but sold the rights to others. He sold the patent rights to his wheel to Moses Hoit and Charles Rogers for $1500 "to make, use and vend within the states of Maryland, Virginia, North and South Carolina, Kentucky, Tennessee, Mississippi, and Georgia and not elsewhere according to Parrish's geography."[7] He sold the rights to various parts of New York and Pennsylvania to others. We saw evidence of Daniel Read doing likewise; and while Amos Miner did engage in the business of making accelerating heads, he soon moved on to other pursuits including more inventing.

There is a long tradition of an association between the Skinner patent wheel at the Whaling Museum and Thomas Howland, a spinning wheel maker who died in 1823. Our guess is that Howland had purchased the rights to make this wheel in some part of New England and that this would account for his association with this wheel. The only three examples we are aware of were found in New England.

Another patentee discussed elsewhere is Joel Farnham, who was a prolific maker along with other members of his family. He holds a patent jointly for a spinning wheel; but since he made two unusual ones, we can only speculate that it is the five-legged model with an accelerating wheel (Fig. 6-26) that was the one patented one in 1825.[8] Both his four-legged and his five-legged versions were widely copied. E. B. Sandford (sic) has an 1816 patent for a spinning wheel which is almost certainly the odd upright double flyer wheel of which several signed examples remain.[9]

It is frustrating that there are few marked Connecticut chair wheels, so we have no clue who may have patented it originally. That it was initially a patent wheel seems a safe assumption. The only signed chair wheel with a name is by "J. Miles," and he is not in the patent lists (Fig. 6-11).

H.F. Wheeler patented an "inclined spinner" in 1838, and we have the full particulars on this wheel.[10] We do not know if it was the first spindle wheel designed to obviate the need for the spinner to walk, but we think so. As such, it gave rise to a whole host of successors. Wheeler modestly refined his patent in 1846. An example of this wheel is at the Henry Ford Museum in Dearborn, Michigan.[11] The key feature, which others emulated in a variety of ways, was the arrangement whereby the spindle spinner remained seated while some ingenious contrivance allowed the spinner to move the spindle away from him or her by means of the use of a treadle and counterweight. When the treadle is released, the counterweight brings the spindle back to the spinner. The spinning head moves along two metal rods, which function as guides. Similar wheels to Wheeler's are associated with Mennonite communities in Ohio and Iowa.

In 1866, when a rebirth of home spinning occurred during and immediately after the Civil War, John Green made modest modifications to Wheeler's patents and patented his own version (Figs. 9-1 and 9-2).[12] The dramatic paint and decal work on this wheel certainly is a long way from the utilitarian finishes of spinning wheels in the first half of the 19th century (Fig. 9-3).

Fig. 9-1. While a little bit large for most collectors, this beautifully painted and decaled example of the John Green patent version of an earlier Hiram Wheeler patent is magnificent. Certainly dramatic in a museum setting, but rather long for the living room one wonders where it would have resided in a 19th century home. DW=42 ½". *Courtesy of David Pennington.* $700-750.

Fig. 9-2. A close up shot to indicate some of the complexity of the previous wheel. *Courtesy of David Pennington.*

Fig. 9-3. Another close up of the John Green wheel to show the elaborate paint and decoration. This wheel was not intended for outdoor or barn use! *Courtesy of David Pennington.*

The next "big step" forward in keeping the spindle spinner "seated" was by Jacob Shaw, Jr. of Hinckley, Ohio. Shaw moved away from Wheeler's method for moving the spindle away from the spinner, the spindle on a fixed track, which yielded a very large machine at rest. Shaw used a hinged arm, which he referred to as a "vibrating frame."[13] The movement of this arm to or away from the spinner was the key. In Shaw's case the pivot point was near the floor (Figs. 9-4 and 9-5). We do not know in what numbers they were made, but it is interesting that members of a utopian community initially setup in Bethel, Missouri, which later moved to Aurora, Oregon, somehow became enamored of this style of wheel and used it in both locations. Three examples have turned up in regional antiques shows and a museum in the Ohio/Michigan area, so it is quite possible that Shaw made some for this market.

Fig. 9-4. A nice example of the Jacob Shaw patent wheel. These take up far less room than the Green wheels and would not use much more space at rest than a wool wheel. Of course, with the "vibrating" arm extended, it would require additional clearance for use. DW=42 ½". *Courtesy of David Pennington.* $700-800.

Fig. 9-5. Another extremely nice example of the Jacob Shaw patent wheel. The black pin striping on a red background suggests that someone was merchandising these; however, the utility of such wheels, when used with pencil roving, is proven by their adoption by the members of a utopian society with communities in Bethel, Missouri and Aurora, Oregon. A third example is at Historic Lyme Village in Bellevue, Ohio. DW=42 ½". *Courtesy of Jeanne Asplundh.* $750-850.

Solomon Dell of Strathroy, Ontario, patented a wheel in 1866 which he referred to as a "Lever Spinning Wheel" (Fig. 9-6). It is like the Shaw wheel with its hinged arm which pivots at a point down near the ground as opposed to the "pendulum wheels" we will look at soon. Within three years, at least two American inventors, as well as two Canadian inventors, patented copies of Dell's machine. Given their ungainly size survival rates are not surprisingly pretty low. Dave is either very lucky or unlucky to have one.

The next form of patented moveable spindle machines are large, ungainly, (really hard to photograph) and difficult to get strung up correctly. It is true that with their little wooden carriages on wheels, they are fascinating to watch in operation. David Current in 1850 in Kentucky appears to have been the first to come up with the idea of the carriages on tracks in place of the guiding metal rods used by Wheeler. David Teter and Jesse Byrkit, both in Jefferson County, Iowa in 1865, patented their own versions.[14] Working examples of Teter's version have turned up in Iowa. Unlabelled versions haunt the collections of demented collectors' attics and garages as no spouse will allow one in the house for long (or short). Fully extended they can be eleven feet long or more!

The best-known form of moveable spindle wheel is the so-called "pendulum wheel" of Lyman Wight.[15] Building on the work and terminology of Jacob Shaw, Jr., Wight came up with the "vibrating pendulum." It is found in a number of versions and was made in the thousands during the 1860s and probably earlier. (Someone could write a book just about all the versions of the pendulum wheel and the detailed differences. The American Textile History Museum has a nice collection of many of the variants from both sides of the border.) Anyone who thinks hand spinning died out by the 1820s can quickly see that hand spinning was quite alive in the last half of the 19th century. The earliest versions have a wooden weight filled with "shot." There are three of these versions, one of which was made near Scranton, Pennsylvania, where Lyman Wight was living until a few weeks after he received the patent. These were probably not made by Wight as he had moved to Waitsville, Wisconsin, before the patent date, which is painted on both surviving examples.[16] Perhaps like Skinner and Read he had sold the Pennsylvania rights to others. There is an example at American Textile History Museum which is the only one exactly like the patent drawing. It was collected by Joan Cummer in New England, so it may have been made by someone with rights to some New England territory.

Fig. 9-6. Solomon Dell's "Lever Spinning Wheel" is a patent wheel from Ontario. Canadian patent wheels were patented by province rather than nationally, so tracking them down is a little trickier than with American examples. Patented in 1866 it was influential enough to spawn at least two imitations in the Canadian market. Interesting but too big for all but museums and barn dwellers. DW=36". *Courtesy of David Pennington.* $500-550.

Eventually (1862) Wight sold his rights to Justin Wait and went to work for him as a shop foreman. He and Wait initially made the pendulum wheels much like the ones in the patent, but with a bench table instead of a turned table (Fig. 9-7). About the time Wight left Wait, Wait began to make them with the familiar X-frame and with a metal ball counterweight (Fig. 9-8). In 1866-67 having been enormously successful in the Midwest, Wait allowed M.S. Morehouse of Cape Vincent, New York on the St. Lawrence River to make and distribute them in the east (Fig. 9-10). By then Wight was out of the picture. The ones made by M.S. Morehouse are slightly different from the Wisconsin versions. Some pendulum wheels were outfitted with smaller drive wheels. Some have single groove drive wheels and some have double grooves. A very close copy of the X-frame was made and patented by Lyons and Lucas in 1866 in Ontario, Canada. The earlier Lyman Wight version was copied with very small changes in a patent by George Stiles in 1866 in New Brunswick, Canada. Two other versions of pendulum wheel were patented and made in eastern Canada in the late 1860s, so Lyman Wight had certainly hit on a successful form of spinning wheel. With local carding mills providing a pencil size wool roving suitable for rapid twisting and little drawing required, these wheels could be quite productive.

Fig. 9-7. This pendulum wheel is a "marriage." The base is from Lyman Wight's earliest work in Wisconsin with Justin Wait, his "partner/boss." All the pieces are marked with a "V." The top piece with the metal ball has a "II" stamped on it. The top piece refers to the top cap, which rests on top of the tall vertical support, plus the metal ball and long vibrating pendulum arm. The original top piece would have all been of wood including the ball. The base was originally in blue paint as were the ones made near Scranton, PA. Curiously, another example, quite complete, of this earliest Wisconsin version has the all wooden top piece, but was not painted blue. When so few examples exist, it is hard to be sure of what was happening during this period of transition. DW=40".
Courtesy of Michael Taylor. If it were in paint complete with a wooden ball, $1500-1700. As is, $700-750.

Fig. 9-8. A very nice example of the pendulum wheel made in Wisconsin. Justin Wait with Lyman Wight working as his foreman made thousands.[16] Not scarce in the Wisconsin area, these wheels are interesting to collectors and fun to spin on. The trick to adjusting the tension between the drive wheel and the accelerating wheel set in the top piece is to rotate the top piece slightly and set it in place with the wooden thumbscrew at the top. DW=43". *Courtesy of James Munsie*. $750-800.

Fig. 9-9. A close up of the spindle assembly on the vibrating arm. These pieces are often lost. They slide on a tapered arm and are held on by the drive cord. Once the drive cord is removed, the odds of the little spindle assembly falling off are pretty high. Tension between the accelerating wheel and the spindle employs a little wedge which holds the spindle assembly. *Courtesy of James Munsie.*

Fig. 9-10. This pendulum wheel belonged to Edna Blackburn, a famous spinner and dyer from Canada. The somewhat smaller drive wheel is typical of those pendulum wheels made by M.S. Morehouse of Cape Vincent, New York who held a license from J. Wait after 1866. However, a virtually identical wheel was patented in Ontario, Canada in 1866 by Lyons and Lucas. DW=42 ½" *Courtesy of Jeanne Asplundh*. $750-800.

Another quite successful patent wheel from this post-Civil War period was the so-called "accordion" wheel. In 1868 James Johnson and J. Wilson Foust of Evansburg, Pennsylvania would appear to be the first to patent what they refer to as "contracting and expanding-frame, composed of two series of toggle-arms, joined together by toggle-joints."[17] One picture from Tennessee of what is probably their wheel exists.[18] Much more successful was the work of Walter B. Walker of Cherokee Station, Kansas who patented a similar wheel in 1874. In the patent application itself, he assigned half his rights to T.D. Brown who made and sold them both in this country and in Canada where Brown took out a patent in 1883.[19] Several of these have surfaced, most with the Canadian date of 1883 (Figs. 9-11 and 9-12). All surviving examples resemble the Canadian patent drawing in having the accelerating wheel next to the drive wheel rather than at the other end of the "accordion" as shown in the U.S. patent. Ironically, that wheel arrangement was the one shown in the Johnson and Foust patent of 1868.

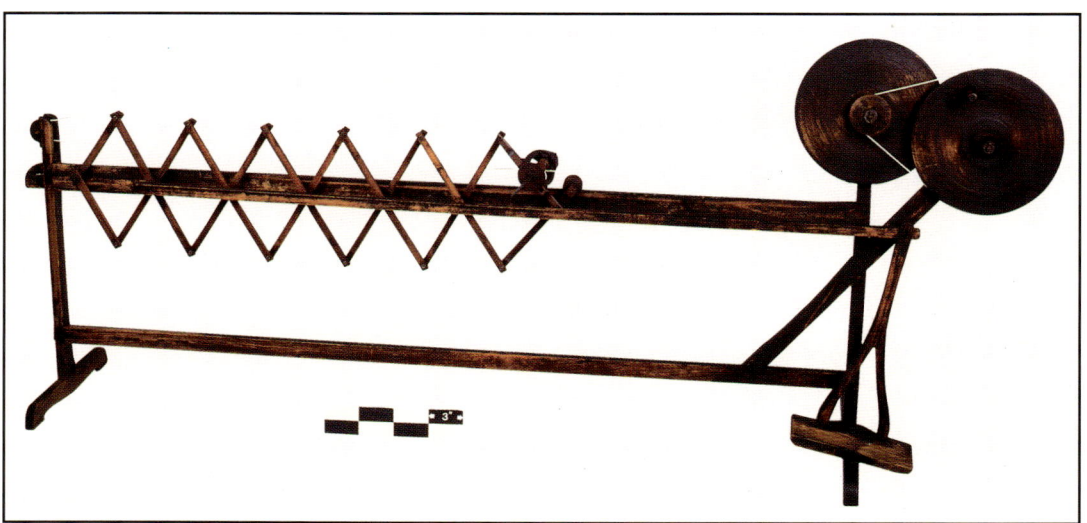

Fig. 9-11. This is a great example of the T. D. Brown "accordion" wheel with the Canadian patent date handwritten. The spindle assembly had to be restored based on the wheel in Fig. 9-12, and it needed a lot of careful adjusting as there are many places for misalignments to occur with use. Interesting, when the wheel was totally disassembled, all the pieces had penciled numbers which suggested that these were broken down for shipping. Scarcity, desirability and a careful restoration based on another complete example make the restorations unimportant to the wheel's value in this case. DW=11 ½". *Courtesy of Jeanne Asplundh.* $750-800.

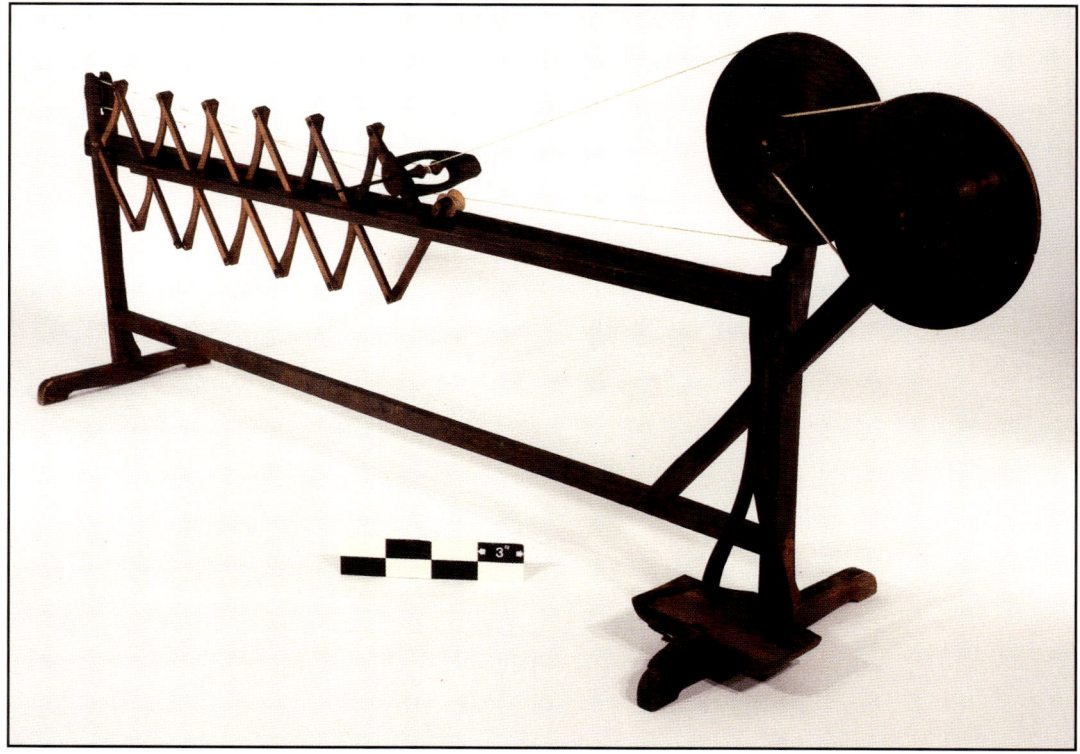

Fig. 9-12. This T. D. Brown wheel also showed a great deal of wear and tear which suggests that these wheels were not just novelties. They are scarcer than pendulum wheels, but their size limits their collectability. They are much harder to put in good spinning order. DW= 12". *Courtesy of David Pennington.* $750-800.

Fig. 9-13. Chelton Matheny took out a patent for this type of wheel in the early 1860's, and it was copied by the aptly named Canadian, Moses Doolittle, who "did little" other than copy poor Chelton Matheny. It is a beautiful piece of work. DW=23 ½". *Courtesy of Jeanne Asplundh.* $800-850.

Chelton Matheny was one of the more persistent of spinning wheel inventors pursuing the moveable spindle principle with three patents in the 1860s.[20] We have seen swinging arms supported from below and from above. Matheny's first patent (1860) has an arm which moved horizontally rather than vertically (Fig. 9-13). At least one of these has survived. The other two were elaborations on the wheeled cart and the pendulum wheels, but so far no survivals exist to show that these were ever made. The 1860 version was at least popular enough that Moses Doolittle in Ontario, Canada took out a patent for a very similar wheel in 1866.[21] We aren't sure if this example (Fig. 9-13) is from George Potts or Walter Guthrie, both of whom made Doolittle patent wheels. It would be nice if this were from Matheny himself, but given his two other patents, he probably was in the idea business rather than the production department.

All of the moveable spindle machines were quite large and took up room. Obviously that was a problem, so it is no surprise that small wheels would be desirable to some. In 1835 shortly before the Patent Office fire an L. Norcross took out a patent on a table model wheel which was essentially a toothed gear which directly engaged a spindle.[22] The gear was turned by a rod with a handle. All that remains is the diagram. In the 1860s when a host of patents for spinning wheels were taken out due to the shortage of manufactured woolen goods during the Civil War and subsequent disagreements with Great Britain, table model spindle wheels became quite popular. They are frequently confused with quill winders, and from period photographs we know they were even used as such, but they were patented as spinning wheels.

In the summer of 1866 three wheel patents were granted, but of the three only that of Wilson and Fairbanks seems to have caught on.[23] Two slightly different variations were produced. The one shown has the spindle support post at the end of the table (Figs. 9-14 and 9-15. Others have the spindle holder fastened to the uprights. These have been found in Missouri and western Illinois near where the patent was taken out (Adams Co., Illinois).

Fig. 9-15. The other side of the wheel shown in Fig. 9-14. The spinner turned the crank with the right hand and spun with the left facing the wheels end on. *Courtesy of Michael Taylor.*

Fig. 9-14. Easily mistaken for a quill or bobbin winder, this Wilson and Fairbanks patent wheel was for those with little space. There are two very different types. This one has the spindle assembly mounted at the end of the table whereas some have it mounted directly below the wheels. This wheel originally came from Hannibal, Missouri right across the Mississippi River from where it was patented. DW=11 ½". *Courtesy of Michael Taylor.* $400-450.

The next patent wheel to make a stir was popular far from where it was patented. J.W. Burkhart of Cameron, Missouri came up with something really original.[24] A number of these have surfaced across Appalachia where they were used as quill winders as well as spinning wheels.[25] Antique dealers in the Staunton, Virginia area reported that they knew where the old factory making these wheels was located, and Mike found several such wheels there. Two different models were produced (Figs. 9-16 and 9-18). Some have little decals on them indicating that it is a "Burkheart (sic) patent wheel" and that the Hawkins Brothers were making and selling them. How these wheels became popular so far from where they were patented remains a mystery. What is original about this wheel is the way tensioning between the accelerating wheel and the spindle is handled.

Fig. 9-16. This example of a Burkhart patent wheel is a magnificent small wheel with great details and original decals indicating the patentee and the retailer. The accelerating wheel is on the opposite side of the solid upright which supports everything. This one is set up the way the patent shows. The one in Fig. 9-18 has the accelerating wheel on the same side. The tensioning system between the spindle and the accelerating wheel is a really original idea. Exceedingly rare example of a desirable wheel. DW=11 ½". *Courtesy of Michael Taylor.* $550-600.

Fig. 9-17. The other side of Fig. 9-16 showing the position of the accelerating wheel. *Courtesy of Michael Taylor.*

Below:
Fig. 9-18. A very nice example of the type of Burkhart wheel made in quantity in the Staunton, Virginia area. The accelerating wheel is on the same side as the drive wheel. Of all the small patent wheels the Burkharts are the most desirable. They are innovative, spin well and are nice to look at! DW=11 ½". *Courtesy of Jeanne Asplundh.* $400-450.

Fig. 9-19. Not a patent wheel but almost certainly a takeoff of the Burkhart wheels, this accelerated table top wool wheel has been found in the Virginia/Tennessee area. It is almost certainly from the 1860's. *Courtesy of Jeanne Asplundh*. $300-350.

Fig. 9-20. The other side of the preceding wheel clearly showing the wing nuts which tighten on the sliding tables by which the tension on the drive cords is adjusted. *Courtesy of Jeanne Asplundh*.

The small wool wheel shown in Fig. 9-19 and 9-20 is not a patent wheel, but it is clearly based on this group of table top wool spinners in the 1860s. It seems to be associated with Virginia and Tennessee which was where the Burkhart wheel was so popular.

Pirating one another's ideas even if there was an existing patent in place was common. W.H. Main of Marietta, Wisconsin patented a clever little spinner in June, 1870. According to an article in *The Spinning Wheel Sleuth* there were two versions of this wheel in production.[26] Their cast metal wheels are clearly marked "W. H. Mains" even though his name on the patent is Main. Apparently with their metal teeth engaging a metal worm gear as a power source, they were quite noisy. That, at least, was a sufficient reason in John Bryce's mind to apply for a patent on a very similar design.[27] His improvement was to employ a leather friction pad against which the metal wheel is pressed. In the patent Bryce shows a metal accelerating wheel, but those manufactured had a wooden accelerating wheel which may have been cheaper to make. Interestingly enough Bryce worked for Justin Wait in Wisconsin. Wait was the entrepreneur who popularized Wight's pendulum wheel. Wait moved to Grand Haven, Michigan after a disastrous fire and Bryce accompanied him.

It was while living in Grand Haven that Bryce received his patent on September 24, 1872. His cast drive wheels have his name, "J. Bryce" and the patent date on them (Fig. 9-21). A number of his little wheels have survived, and if not too rusted work quite well. He was aware of the Canadian market and patented his wheel there in 1873.[28] Apparently, he had someone there making them as two variants appear that come from the Canadian market. The variant shown looks identical to its U.S. cousin, but the cast wheel reads "J. Bruce," not "J. Bryce," and the patent date cast into the wheel is 1873, not 1872 (Fig. 9-22). The remains of another version are shown in Judith Buxton-Keenlyside's book on Canadian wheels.[29] While not complete, it is clear that in this version, the friction pad approach to propelling the accelerating wheel has given way to a drive band of some sort. Whether this was a home conversion or was actually produced in limited numbers is unclear.

Fig. 9-21. The cast iron drive wheel on these has "J. Bryce" and the patent date in raised letters. While patented late (Sept. 24, 1872 in the U.S.), they were popular enough that they turn up occasionally even though few dealers realize that they are spinning wheels. DW=9 ¾" *Courtesy of David Pennington*. $350 if complete.

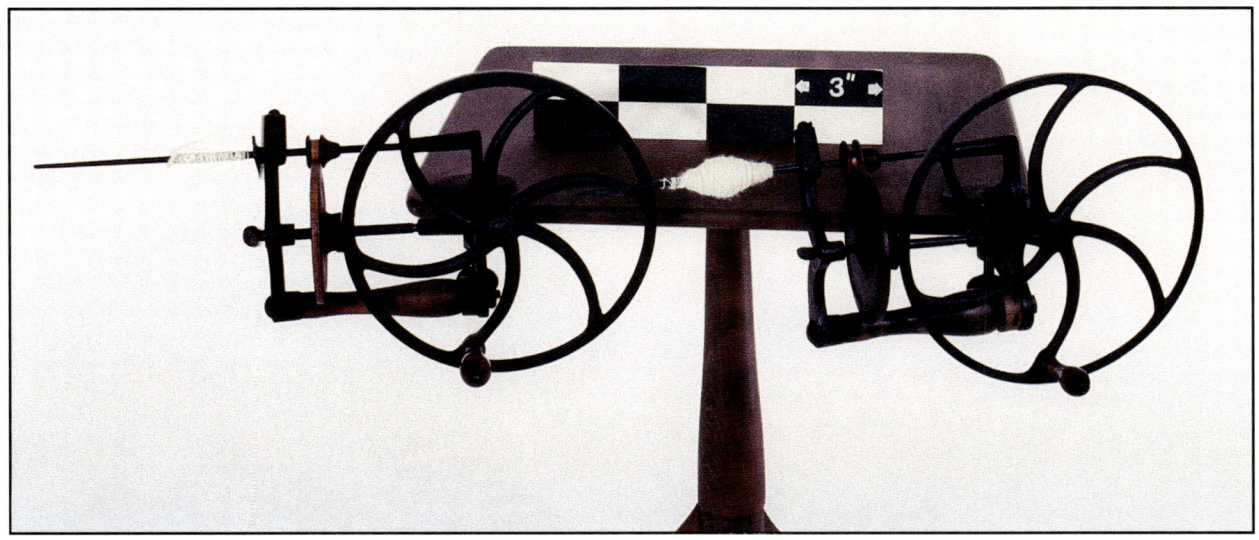

Fig. 9-22. A Bryce and a Bruce! Bryce took out a Canadian patent in 1873 and apparently some were made there. Unless you look at them very closely, you can't tell them apart. The Canadian version has the name "J. Bruce" and the date is "Sept. 24, 1873", not 1872. In both cases Sept. 24 is used, which may be coincidence or laziness. The Canadian version was found in New Hampshire. Both DW's=9 ¾. *Courtesy of Michael Taylor*. $350 each.

In 1871 a table model spindle wheel, the American Spinner, was patented in Maine by George Hathorn.[30] It incorporates a reel with a counting mechanism, and by reassembling the pieces and attaching a metal guide, the reel can easily remove and count the amount of yarn from the spindle (Fig. 9-23). These occasionally turn up, and the one shown has the number 4109 stamped on its various parts suggesting that thousands were made. If so, few survived as they are not commonly found. Here we have a case of a U.S. patent based on a Canadian precursor.

Fig. 9-23. This example of a George Hathorn spinner is in wonderful condition and still has the label with directions on it. The label calls this an American Spinner #7, but it looks exactly like the patent, so we aren't sure why the #7. Each piece is number 4109 which suggests that thousands were produced and they are found in New England, but not commonly. Hathorn, working in Maine, was almost certainly borrowing from John Henry Nute of New Glasgow, Nova Scotia. As shown in this photo, it is a spinning wheel and a swift. There is a metal guide to a make it easy to wind the spun yarn onto the reel. There are holes lined up such that the "windmill" unit may be laid on its side and function as a reel. DW=21 ¼". *Courtesy of David Pennington.* $400-450.

John Henry Nute of Nova Scotia had patented his famous "hurdy" wheel in 1870. These were very successful and many examples have survived (Figs. 9-24 and 9-25). With their brightly painted red wheel with black dots they are quite dramatic. Having many easily lost parts, it can be hard to find a really complete one. Perhaps, they were introduced into Maine quickly where Hathorn saw one and came up with his own version. In 1874 Lucius Bishop of Nova Scotia patented a wheel similar to Hathorn's (Fig. 9-26).[31] Hathorn and Bishop were not the only ones with the right angle table-top idea. The one shown in Fig. 9-27 has not been identified by patent or label, but certainly belongs in this Nova Scotia/Maine patentor's group. Judging by the number of "hurdies" surviving, Nute was by far the most successful of the three inventors.

In 1880 Seraphin Vigeant and Pierre Desmarets of Quebec, Canada patented a virtually all-metal saxony, and at least three of them have survived. Not nearly as successful as Nute's table top, multiple function wheel, they are still quite dramatic and approach the realm of fine art sculpture.[32]

Why patent spinning wheels in the late 19th century? This has been asked both formally and informally by spinning wheel collectors and historians.[33] Many people assume that with the coming of the Industrial Revolution in the early 19th century only rustics living far from settlements would continue to spin. The answer to the question is complex. In part, spinning and weaving had become symbols of women's work and domestic self-sufficiency.[34] Transportation was arduous with no cars and bad roads. In part, many people did live in the country and some raised sheep. They had access to a key raw material, wool fleeces. Perhaps, most importantly, they had access to a local carding mill which could quickly and easily turn a dirty fleece into carded wool and would take a part of the product in exchange for the service. Joel Farnham and his family ran a carding mill. According to family records they continued to produce pencil roving right up to the end of the 19th century.[35] Making and selling wheels might have been a sideline that gave their customers a key tool and would encourage them to use the carding service. One key piece of evidence is

Fig. 9-24. A very nice example of a "Nute's Combined Spins". In most cases they would simply have been clamped to a work table, but this little table is cut out to give additional stability. This little wheel can be used as a spinning wheel and counting reel. They are referred to as "hurdy" wheels and must have been extremely popular as one collector remembers seeing a single dealer with dozens of them in the 1970's. The red wheel with black dots is quite lively. DW=18". *Courtesy of Jeanne Asplundh.* $300-350 with the custom made table.

Fig. 9-25. This Canadian "hurdy" wheel patented in 1870 was extremely popular. Parts to these are often missing, but a complete one is a nice addition to a collection. DW=18". *Courtesy of David Pennington*. $200-250.

Fig. 9-26. A very nice example of a Lucius Bishop table-top wheel. Bishop of New Minas, Nova Scotia patented his wheel in 1874. With their right angle turn for the drive cord, they are quite dramatic and exceedingly rare. DW=25". *Courtesy of David Pennington*. $350-400.

Fig. 9-27. A cute little example of the Bishop patent idea but with a solid wheel. It might easily be thought of as a quiller if it were not for the existence of the Nute, Hathorn, and Bishop patent wheels. DW=13". *Courtesy of Jeanne Asplundh*. $150-200.

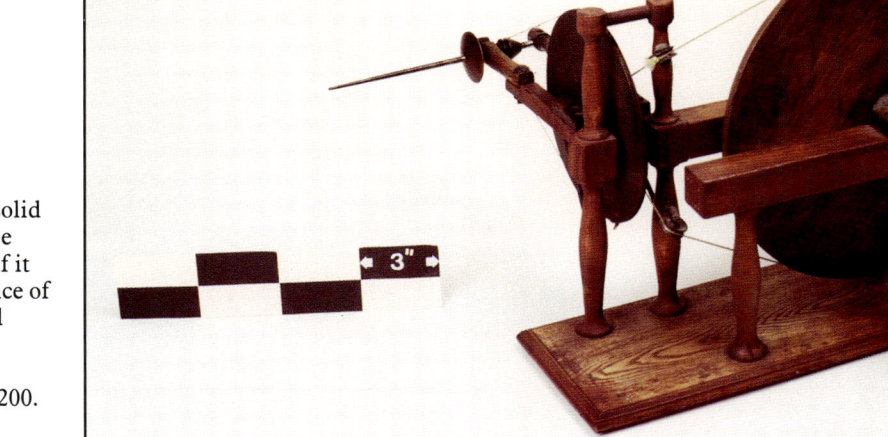

an old postcard showing Mrs. Elizabeth Kreger of Liberty (Tioga County), Pennsylvania spinning on her pendulum wheel around 1901.[36] In her left hand is pencil roving, which is a form of carded wool from a carding mill. Pencil roving does not need to be "drafted," i.e. pulled apart by the spinner, as it is spun. You just turn the drive wheel and depress the treadle on your patent wheel and make yarn by the mile.[37] So, the various accordion wheels, track wheels, pendulum wheels, lever arm wheels came into their own during and after the Civil War when woolen goods were scarce or expensive, and the little local carding mills stayed open. Rural women, cultivating their "distaff" image and producing something that could be traded at the general store, were then a ready market for such patent wheels.

As people rummage through their attics looking for e-Bay fodder or the chance to appear on *The Antiques Roadshow*, some other patent wheels, which were more than a hope and a piece of paper, may turn up. Or as curators of local historical societies and small museums look through documents in their files, another lost patent drawing and description like those of Skinner and Read may turn up, but for now that is most of what is out there. What a rich source of information was lost in the Patent Office fire of 1836!

---

[1] D. Demming, "Account of the New York Accelerating-Head for Spinning Wool," *Archives of Useful Knowledge* Vol. II, No. 2 (October, 1811): 105.

[2] M.D. Leggett, comp. *Subject-Matter Index of Patents for Inventions Issued by the U.S. Patent Office from 1790 to 1873, Inclusive.* (Washington: Government Printing Office, 1874): 1393-1401.

[3] D. Demming, 104. It looks exactly like the standard model of which hundreds of thousands were made over the next 80+ years. Frank White, "Heads Were Spinning: The Significance of the Patent Accelerating Spinning Wheel Head," *Annual Proceedings of the Dublin Seminar.* (Boston: 1999) pp. 64-81.

[4] Craig F. Evans, "Elijah Skinner and Thomas Howland," *The Spinning Wheel Sleuth* 21 (July 1998): 2-5 and Michael B. Taylor, "The Pleasant Spinner," *The Spinning Wheel Sleuth* 27 (January 2000): 6-8. In both cases curators working their way through old files found lost patent descriptions.

[5] Patricia Baines, *Spinning Wheels, Spinners and Spinning*, (New York: Charles Scribner's Sons, 1977): 146-149. Baines survey of the European sources does not turn up any examples of accelerated wheels where the connection is using a drive cord or band. However, the accelerating principle is found by 1688 in the description Randle Holme gives of a "girdle wheel." Randle Holme, "Extracts from the Academy of Armory by Randle Holme of Chester, 1688," Excerpted by Alan Raistrick. *The Spinning Wheel Sleuth* 20 (April 1998): 4.

[6] Craig F. Evans, "Elijah Skinner and Thomas Howland," *The Spinning Wheel Sleuth* 21 (July 1998): 2-5.

[7] Ibid., 3.

[8] Leggett, 1400.

[9] Ibid., 1400.

[10] Patent Number 710, U.S. Patent Office, Washington D.C.

[11] Patent Number 4892, U.S. Patent Office, Washington D.C.

[12] Patent Number 59,210, U.S. Patent Office, Washington D.C.

[13] Patent Number 5847, U.S. Patent Office, Washington D.C.

[14] Patent Numbers 7614, 47,685 and 50,094, U.S. Patent Office, Washington D.C. Current's patent may have run out by then.

[15] Patent Number 14,482, U.S. Patent Office, Washington D.C.

[16] Victor L. Hilts, and Patricia A. Hilts. "Not for Pioneers Only: The Story of Wisconsin's Spinning Wheels," *Wisconsin Magazine of History* (Autumn 1982): 3-24 is one of the seminal articles about patent spinning wheels. A must read for anyone interested in American spinning wheels. Also, see Michael Holcomb, "A Lyman Wight Pendulum Wheel," *The Spinning Wheel Sleuth* 23 (January 1999): 4-5, for a picture of one of those that were made in the Scranton, Pennsylvania area. They are in blue paint with black trim and red lettering. The earliest version of the Lyman Wight pendulum wheel is shown in Joan W. Cummer, *A Book of Spinning Wheels*, (Portsmouth, N.H.: Peter J. Randall, 1993): 54-55.

[17] Patent Number 83,970, U.S. Patent Office, Washington D.C.

[18] Sadye T. Wilson, and Doris F. Kennedy, *Of Coverlets: the legacies, the weavers.* (Nashville, Tennessee: Tunstede, 1983): 421.

[19] Florence Feldman-Wood, "A T.D. Brown Accordion-Arm Wheel," *The Spinning Wheel Sleuth* 23 (January 1999): 11-12.

[20] Patent Numbers 27,059, 62,351, and 96,937, U.S. Patent Office, Washington D.C.

[21] Grant Betzner, "The George Potts Spinning-Wheel Shop and the Moses Doolittle Wheel," *The Spinning Wheel Sleuth* 24 (April 1999): 8-9.

[22] Leggett, 1393. A drawing of Norcross's "accelerated spinner" survived the fire or he reapplied and was given patent number 8899X, but no written description survives.

[23] Jane Lenderman-Kruse and Don Kruse, "Tabletop Patented Spinning Wheels," *The Spinning Wheel Sleuth* 19 (January 1998): 4-5.

[24] Patent Number 81,594, U.S. Patent Office, Washington D.C.

[25] Sadye T. Wilson, and Doris F. Kennedy. *Of Coverlets: the legacies, the weavers.* Nashville, Tennessee: Tunstede, 1983: 76-77 shows two late 19th century pictures of Laodicea "Aunt Dicie" Fletcher, a professional weaver from Rugby, Tennessee, using a Burkhart spinner as a quiller. John Rice Irwin collected one locally for his Museum of Appalachia, Norris Tennessee.

[26] Jane Lenderman-Kruse and Don Kruse, "Tabletop Patented Spinning Wheels," 2-3.

[27] Ibid., 3-4. see also Hilts and Hilts, "Not for Pioneers Only."

[28] Judith Buxton-Keenlyside, *Selected Canadian Spinning Wheels in Perspective: An Analytical Approach*, (Ottawa: National Museums of Canada, 1980): 255.

[29] Ibid., 258.

[30] Patent Number 121,517, U.S. Patent Office, Washington D.C.

[31] Nelda Davis, "Bishop Tabletop Patent Wheel," *The Spinning Wheel Sleuth* 26 (October 1999): 8-10.

[32] Pictures two of these are in Judith Buxton-Keenlyside, *Selected Canadian Spinning Wheels in Perspective: An Analytical Approach*, 165 and Joan W. Cummer, *A Book of Spinning Wheels*, 213.

[33] Victor L. Hilts and Patricia A. Hilts, "Why Patent Spinning Wheels: Some Additional Thoughts," *Spin-Off, The Magazine for Handspinners* (Spring 1996): 90-95.

[34] Laurel Thatcher Ulrich, *The Age of Homespun: Objects and Stories in the Creation of an American Myth*, (New York: Alfred A. Knopf, 2001): 17-18.

[35] Pamela Goddard, "Farnham Family Textile Tools," *The Spinning Wheel Sleuth* 10 (October 1995): 6.

[36] Hilts and Hilts, 95.

[37] Michael Taylor came to this realization when he happened upon some pencil roving and began to spin it up on his pendulum wheel as an experiment. Years ago Dave Pennington had mentioned watching a demonstrator at Greenfield Village in Dearborn, Michigan spinning on a wool wheel with incredible rapidity using pencil roving and wondering how authentic that was. Turns out it was very authentic.

Chapter 10
Famous Makers

There are two great wheel-making families examined here: the Sanfords of Newtown, Connecticut, and the Farnhams of "near Owego," New York. We will look at Beardsley Sanford, a Sanford cousin, who learned his trade in Connecticut and then moved to Fergusonville, New York, where he made a wide variety of exceptionally well-crafted wheels. We will also look at Alpheus Webster of Green County, New York, who patented and made two extraordinary wheel types, and Daniel Danner of Manheim, Pennsylvania, who made gorgeous Irish "cassel" wheels and other styles as well. Part of the reason for looking at these folks is that they left behind signed examples of their work and records. No one made better looking wheels than John Burley or W. Fancher; but their surviving output is miniscule, and we know little about them.

The lineage of the various Sanfords making double-flyer wheels is traced in the chapter on double-flyer wheels. Beardsley Sanford, on whom we will concentrate our attention here, moved to Fergusonville, New York, after learning his trade from his cousins.[1] He was born in New Haven, Connecticut in 1790 making him almost an exact contemporary of Elias Bristol Sanford and Josiah Sanford. He died in Fergusonville, New York, in 1868. We believe he learned his trade from either Isaac Sanford, father of Elias Bristol and half brother of Josiah, or from Samuel Sanford, Isaac and Josiah's father and the clan patriarch. Beardsley made a double-flyer wheel very similar to one marked "I. Sanford" and one marked "S.S." (Fig. 10-1).[2] He used a very dramatic stamp to mark his wheels and reels (Fig. 10-2).

Fig. 10-1. This signed "B. Sanford" double is very similar to some of those of "I. Sanford" and his half-brother, Josiah Sanford. Beardsley Sanford was a cousin of Isaac Sanford, but closer in age to Josiah Sanford. Elias Bristol Sanford was Isaac's son, and he signed his wheels "E.B. Sanford". *Courtesy of David Pennington.* $500-550.

Fig. 10-2. The mark "B. Sanford" used by Beardsley Sanford. On later pieces one can see where the stamp has broken by looking at the "O". In some instances the stamp looks like "D. Sanford", but the identical broken "O" and slightly smaller "D" suggests that the "B" also broke and then looked like a "D". *Courtesy of David Pennington.*

There are wool wheels, saxonies, double-flyer wheels, Alpheus Webster-like double treadle, double wheels (Figs. 10-4 and 6-6), and reels with the "B. Sanford" mark. All are wonderfully made. The reel shown here (Fig. 10-3) also answers the question of who the "D. Sanford" mark was for. It is clear from looking closely at the two marks on this reel that the "B. Sanford" stamp broke, and that when Beardsley used it in this condition, if he wasn't careful, the "B" looked like a "D."[3] The Webster-like double treadle, double wheel flax wheels make sense because Alpheus Webster was working in Greene County, New York, which is close to where Beardsley Sanford was working.

Beardsley Sanford's reel has a simple solution to a common problem with reels. If the arms are turned counter-clockwise on most "click" reels, the flexible arm, "slapper," will break or gears will wear. Beardsley's solution is quite simple (see the more complex solutions of Joel Farnham which follow). He put the slapper inside a large metal screw eye, so that it would click in either direction and break in neither.

Fig. 10-3. A reel with two marks on it: "B. Sanford" and "D. Sanford". While Daniel Sanford was Beardsley's father, a careful examination of this mark shows that the B. Sanford stamp had broken and that the "D" is really a "B". The reel is unusual in that the base is somewhat early looking while the turned upright with the clock-like head is almost modern, but it is right as is. The table is painted an old red. The slapper is inside a metal screw eye and can slap in either direction, but break in neither. The gear in the open face has forty teeth and the circumference around the arms is 74" as one would expect from a New York maker. *Courtesy of Michael Taylor.* $450-500.

Fig. 10-4. An absolutely fabulous "B. Sanford" double treadle, double wheel flax wheel of the sort likely patented by Alpheus Webster in 1812. Webster's patent would have run out by the 1830's when Sanford would have been making these. Sanford used a standard wooden rim on his drive wheel unlike the metal versions favored by Webster. Sanford did go to the trouble of making his threaded rod for the tension device in the same painstaking way which Alpheus Webster had initially. Sanford painted the rim black, which is unusual in a sectioned wheel. The accelerating wheel is solid and painted black. See Fig. 6-6 for its twin. DW=14". *Courtesy of David Pennington.* $800-850.

The Farnham clan is another group of wheel-makers of great skill and ingenuity about whom we know a fair amount. Joel Farnham moved to Tioga County, New York in 1794. He began making wheels soon after arriving and probably took on Enoch Slosson Williams, who was his wife's cousin, and Jesse Trusdell as apprentices or helpers.[4] Many wheels of a variety of styles marked "J. Farnham," "J. Farnham, *Owego*," and "J. Farnham *near Owego*" have been seen. His wool wheels and saxonies are not remarkable, except that the wool wheels have eight grooves in the wheel rim (Figs. 10-5, 10-6, and 10-7). Later on his sons, Joel, Jr., and Frederick Augustus, made wheels with their own marks. It is reported that another son, Charles, also made wheels.

Fig. 10-6. A signed "J. Farnham" Miner's head with the wheel painted green. *Courtesy of James Munsie.* $100-125.

Fig. 10-5. A signed "J. Farnham" wool wheel with a drive wheel having eight grooves in it. Bill Ralph who restored many Farnham wheels and lived near where the Farnhams worked wrote that all "J. Farnham" wool wheels had eight grooves. Some of them come with signed Farnham Minor's heads, which enhances their value greatly. DW=46". *Courtesy of David Pennington.* $650-700 with a regular Minors' head.

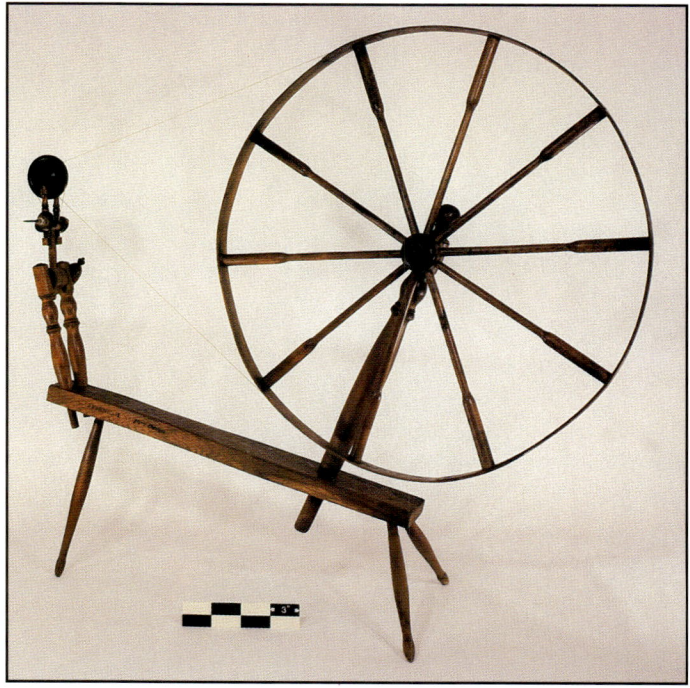

Fig. 10-7. Notice the difference between the drive wheel in this "J. Farnham" and the previous one. This one has an old, but married, drive wheel. However, it has a "J. Farnham" marked Minor's head with the accelerating head painted green. Very hard to value such a piece. DW=43". *Courtesy of James Munsie.* $300-400.

The Farnham clan is of special interest because of 1] their four-legged flax wheels (Fig. 4-12), 2] J. Farnham's five-legged flax wheel (Fig. 6-26), and 3] a patent issued to A. Wittse and J. Farnham of Tioga, New York in 1825.[5] The four-legged flax wheels are distinctive and represent a specific type of American flax wheel (Fig. 10-8). Examples with a variety of marks, including "J. Farnham" (Fig. 10-9), "J. Farnham Jr.," "F.A. Farnham" (Fig. 10-10), "E.S. Williams," and "J. Truesdell," are attributable to this group. Enoch Slosson Williams and Jesse Truesdell were working in nearby Newark Valley, New York. Truesdell also used the mark "J.T."[6] His wool wheels and those of E.S. Williams are somewhat different than those of J. Farnham in having fewer grooves in the wheel (5) and a different rotating block with two holes for an accelerating head rather than a rotating barrel (Fig. 2-29). It is not clear why there are two holes rather than one. Truesdell is said to have taken over Williams shop in Newark Valley.

Fig. 10-8. A spectacular Joel Farnham four-legged type flax wheel. The tension device is the horizontal threaded rod. Only the "J. Farnham" versions have the place for the extra bobbin, and this example is the only complete one with proper extra bobbin and distinctive distaff. DW=20". *Courtesy of James Munsie.* $800-850.

Fig. 10-9. The "J. Farnham" mark on a four-legged type. Sometimes as here "*Owego*" is added and in some cases "*near Owego*" is added. Since Farnham was actually working in Tioga, New York, it seems likely that these wheels were being marketed far enough away that people might know where Owego was, but not the smaller city of Tioga. DW= 20". *Courtesy of Jeanne Asplundh*. $700-750.

Fig. 10-10. A nearly identical wheel but signed "F.A. Farnham". Frederick Augustus Farnham was one of Joel's sons involved in the enterprise who made wheels on his own along with Joel Jr. DW= 20". *Courtesy of Jeanne Asplundh*. $700-750.

Fig. 10-11. A wonderful example of a "J. Farnham" reel. The inventiveness of Joel is seen not only in the two unusual wheels he made but also in his two attempts to make a better reel. The problem reel owners faced was that if a child turned the reel counterclockwise long enough they would break the slapper or harm the gears. (It still happens.) Farnham's solution on this winder was to thread the exposed trip arm; so that if the reel is turned counterclockwise, the trip arm just rotates out of the way. *Courtesy of James Munsie.* $400-450.

When we first noted the patent given to Wittse and Farnham in 1825, we immediately assumed it was for the four-legged flax wheels. Then two five-legged solid wheeled, double treadle, double wheels marked "J. Farnham" turned up (Fig. 6-28).[7] Two signed examples have surfaced, one in a museum in Milan, Ohio and one in the Home Textile Tool Museum in Orwell, Pennsylvania (mailing address: RR1, Box 141A, Rome, PA 18837). While there are many makers' marks on the four-legged wheels ("C. Heese," "Schoonover" and "L. Brown" in addition to the Farnhams, Williams and Truesdell), only the "J. Farnham" mark appears on any of the five-legged models. So the patent issue remains a mystery. His inventiveness shows up in two different mechanisms he developed for keeping the slappers on reels from breaking (Fig. 10-11). One way involved threading the rotating trip arm on the axle such that if the reel is turned counter-clockwise (the wrong way) the arm "unthreads" and eventually moves out of the way. A reel marked "F.A. Farnham" also employs this approach. Another Farnham solution was seen and described by Bill Ralph in *The Spinning Wheel Sleuth*[8]. The principle is shown on a non-Farnham reel shown in Figs. 14-54, 14-55, and 14-56.

The stories of the Sanford's and the Farnham's suggest that spinning wheel manufacturing was a trade handed down from generation to generation, and that others were brought into the trade by marriage. Those employed over time, not surprisingly, began to see recurring problems and to invent solutions or improvements. Joel Farnham's patent comes 30 years after he starts making wheels. Elias Bristol Sanford is a third generation spinning wheel maker (for all we know his grandfather, Samuel, learned from a father or uncle), and Beardsley Sanford probably learned his trade in Samuel's shop working alongside the whole clan. Certainly both Elias Bristol and Beardsley were familiar with problems in the basic Connecticut designs for double-flyer wheels and reels. The shop records of wheel makers show that they frequently did repairs. These repairs pinpoint problem areas, places where there were opportunities for innovators.

The story of Alpheus Webster as an inventor is told in the chapter on patent wheels. He made two extraordinary types of wheels, one of which fizzled, and one of which was a regional success. His double treadle, single wheel never caught on and three examples are known, all marked "A. Webster." His double treadle, double wheel was more successful, and we have shown the variations in the chapter on unusual wheels, and here because Beardsley Sanford was one of the copiers. They lived in adjacent counties in New York, and since patents were only good for 14 years, it is possible that Beardsley, who was about 20 when Alpheus received his patents, either received permission to make this type of wheel or did not need it.

Daniel Danner was a gifted turner and made many spectacular wheels and reels (Fig.10-12). His work is also discussed in the chapters on Irish castle wheels and eastern Pennsylvania saxonies.[9] His Irish castle wheels and his reels show some simplifications in turnings over time which suggests that he may have needed glasses or that someone less talented was helping. There is no indication that he charged less for the simpler turnings. Gabriel Schaffner worked in his shop from 1835-1838 which were prime years for Danner. Danner sent a load of 20 "cassel" wheels to a John Erb in Carlisle, Pennsylvania in 1837. He had to take two back and place them with another merchant, but that is still an astonishing number of cassel wheels for one order. In 1839 he made 42 wheels, but only 16 wheels in 1840. The paper labels on his wheels seldom survive and those that do are tantalizing as they include both dates and numbers, but seldom are fully legible (Fig. 5-5). Bill Leinbach reports having seen a Danner Irish castle dated 1840 with the number 693. Danner started making wheels by 1824. If his typical output in his early years was around 45 wheels of all kinds per year, 693 would not be too high to be believable. On February 25th, 1858 he finished four cassel wheels. A reel with his name dated 1868 is at The Hershey Museum. So even if Danner tapers off his wheel production to about 18 a year after 1840, he would have made another 500 wheels and about 1200 over his working career. His son, Aaron, worked in the shop in the late 1850s. The base of an umbrella swift with Aaron's mark exists although no signed wheels by Aaron are known.

Fig.10-12. A Daniel Danner Irish "cassel wheel" and five armed reel. Danner did not normally sell them in pairs, but it did happen. Having both is great. The reels are much scarcer than the cassel wheels, which seem to have been kept as family heirlooms. The labels on the reels were part of the dials and are generally in good condition whereas the paper labels on the rear legs of the Danner cassel wheels are almost gone or illegible. DW = 17 ½". *Courtesy of Michael Taylor.* Cassel wheel $2000-2200. Reel $750-850.

Danner's reels changed over time. He went from a drop-arm release mechanism for removing skeins in 1825 to the more simple and typical solution of not putting a lip on one of the arms in 1828.[10] The most spectacular thing about his reels is that they are five-armed.

Danner seems to have been well aware of what would make his work distinctive and desirable in the market. His saxonies closely follow the pattern set by Samuel Humes (Figs. 3-3 and 11-2), but he adds the reeling pin feature. Danner's drive wheels have four equal parts instead of the more typical two long and two short. His ongoing production of Irish castle wheels suggests that he had a "high end" model for bridal gifts. Unfortunately we do not know for sure what his wool wheels looked like although his account books show him making them.

---

[1] In the second printing of *A Pictorial Guide to American Spinning Wheels*, which we did in 1977, we corrected some obvious mistakes we had made in the first printing. A descendant of Beardsley Sanford contacted us, gave us Beardsley's first name, which we didn't have, and made the point that he worked in Fergusonville, New York and not Connecticut, so we dutifully corrected it. Now it turns out that he was from both.

[2] Eliza Leadbeater, *Spinning and Spinning Wheels*, (Shire Album 43, Shire Press, 1979): 18 shows a picture of the "S.S." double-flyer in her collection and Florence, Feldman-Wood in her article, "Double-flyer Spinning Wheels," *The Chronicle of the Early American Industries Association* Vol. 53, No. 4 (December 2000): 134 shows the "I. Sanford" double-flyer in question. Both Isaac Sanford and Josiah Sanford used more than one finial and one spoke pattern.

[3] This is important because a "D. Sanford" wool wheel is listed in Florence Feldman-Wood's, "Spinning-Wheel Maker List: List #3" *The Spinning Wheel Sleuth* (October 2001):19. Beardsley's father's name was Daniel, but as of now there is no reason to think he was a wheel-maker.

[4] A number of excellent articles on the Farnham family and their wheels have appeared in *The Spinning Wheel Sleuth* including Pamela Goddard's "Farnham Family Textile Tools," *The Spinning Wheel Sleuth* 10 (October 1995): 4-6, and Bill Ralph's trio: "How to Identify Farnham Flax Wheels," *The Spinning Wheel Sleuth* 10 (October 1995): 2-3, "Identifying Farnham Great Wheels," *The Spinning Wheel Sleuth* 13 (July 1996): 5-6 and "A Rare Farnham Accelerating Wheel," *The Spinning Wheel Sleuth* 15 (January 1997): 2-3. Joel's wife, Ruth, was the daughter of Enoch Slosson according to an article by L.W. Kingman in John E. Smith, ed., *Our Country and Its People: A descriptive and biographical record of Madison County, New York*, (Boston: The Boston History Company, 1899): 497.

[5] M.D. Leggett, comp. *Subject-Matter Index of Patents for Inventions Issued by the U.S. Patent Office from 1790 to 1873, Inclusive.* (Washington: Government Printing Office, 1874): 1400.

[6] Neil Poppensiek, "Jesse Truesdell of Newark Valley, New York," *The Spinning Wheel Sleuth* 26 (October 1999): 6-7.

[7] Bill Ralph, "A Rare Farnham Accelerating Wheel," *The Spinning Wheel Sleuth* 15 (January 1997): 2-3.

[8] Bill Ralph, "Farnham Reels," *The Spinning Wheel Sleuth* 12 (April 1996): 4.

[9] William A. Leinbach, "Daniel Danner: The Man Behind the Wheel," *The Spinning Wheel Sleuth* 4 (January 1994): 6-7 and James D. McMahon, "Daniel Danner, Woodturner of Manheim, Lancaster County: An Early Nineteenth-Century Rural Craftsman in Central Pennsylvania." Master's Thesis, The Pennsylvania State University at Harrisburg, 1992 are two excellent sources.

[10] Irwin Richman, *Pennsylvania German Arts: more than hearts, parrots, and tulips*, (Atglen, Pa: Schiffer Publishing, Ltd., 2001): 80 shows a picture of this remarkable reel with the drop arm, which is at The Heritage Society Museum of Lancaster County. It is a wonderful piece with a label in excellent condition.

Chapter 11

Eastern Pennsylvania Saxony Makers

Eastern Pennsylvania is a wonderful region for spinning wheel collectors. There are many saxonies with marks on them, which can be correlated with one another and with existing shop records. In eastern Pennsylvania we find three groups of signed wheel makers working in Lancaster County, Berks County, and Bucks County with various borrowing going on. Luckily for us, saxonies in this region were marked and until recently stayed in their area so that collectors could begin to recognize patterns. Wheel makers sometimes dated and numbered wheels.

Daniel Danner is well known because he left marked wheels, and his shop records have been analyzed in James D. McMahon's Master's thesis.[1] At the Hershey Museum there are two of his saxonies, both are pictured in McMahon's thesis. The one with the paper label affixed to the high end of the table is very similar to the one on display at The Heritage Center Museum of Lancaster County.[2] Also, the spokes are similar to the ones in many of his labeled Irish castle wheels. The other saxony at the Hershey Museum is dated 1824 and is a bit of a puzzle. The spoke turnings are like those of another group of eastern Pennsylvania makers. The label is in a different hand and is affixed to the front of the table. The legs and the wheel posts have the "wrong" turnings. The finials on the maidens are from a different group of makers. The only thing Danner-like about the wheel is that the drive wheel is in four equal parts. Mike's guess is that if they pulled the paper label off, the Hershey Museum might find the mark of either "J. Killian" or "P. Killian," but who would take the chance? What would settle the matter less dramatically would be the appearance of another early Danner with a similar label. The wheel was likely collected by George Danner, Daniel's son, who made a museum in his father's memory, which later became the Hershey Museum. You couldn't have a better provenance, but George was putting the collection together after his father died and was probably looking for wheels with his father's paper label on them.[3] The wheel is very close to the work of the Killians (Fig. 11-1). Antique enhancers were a part of the scene in the 1890s too.

Fig. 11-1. Marked "J. Killian", for either Jacob Killian or John Killian, this wheel is in the kind of condition a collector dreams about. It is definitely Lancaster County, Pennsylvania. A similar wheel marked "P. Killian" is in a museum in Iowa! DW=19 ½". *Courtesy of Michael Taylor.* $650-700.

It is also possible that Danner made wheels like Killian's in the early 1820s, and then switched to copying the style of Samuel Humes, who may have been cutting back production by 1827 after nearly of 50 years of work as a turner (Fig. 11-2). The two later Danner saxonies, with labels and construction like his Irish castles, are virtual clones of Humes work. The main structural differences are that Danner had a drive wheel with four equal parts and a hole in the table for a reeling pin. From across a room a Danner saxony and a Humes saxony are virtually indistinguishable. Fortunately, Humes seemed to have marked every saxony with "S. Humes" and the date. He occasionally used a "Lancaster" stamp too. Starting around 1817 he began to number his wheels, and the early ones used the symbol § in front of the number. By 1828 Humes is using an "8" in place of the §. He may have also broken the § mark. In 1836 Humes died and his inventory lists parts of wheels and reels on hand which would substantiate the memory of one long-time Lancaster wheel collector who recalls seeing an "S. Humes" wheel marked 1836.[4]

Both Humes (1753-1836) and Danner (1803-1881) worked as turners for a long time. Humes is listed as making chairs in 1775, but his earliest known wheel is 1791. He worked until 1836. Danner, according to shop records and surviving labeled pieces, worked from 1824 until at least 1868. In 1840 Danner made twelve "spinning wheels" (probably saxonies) @ $3.50 each, three "cassel" wheels @ $4.00 each, two "wool" wheels @ $4.00 each and six reels @ $1.25. So far we have seen no labeled wool wheels of Danner's. As a turner he made and repaired a wide variety of items and did metal work as well. Wood seemed to be more valuable than his time as a saxony ($3.50) would take more time and skill to produce than a wool wheel ($4.00). He made everything from brushes to wagon wheels. He repaired flyer hooks regularly, and in one case it appears that it took a spinner only eighteen months to wear through her hooks. He dealt with retail merchants, who paid in cash, and widows, who sometimes paid in spun yarn.

The work of Danner, Humes and the Killians ("J. Killian" and "P. Killian") represent a Lancaster County style of saxony: no wheel support posts, flyer teeth on the top right arm of the flyer, and "shotgun shell and olive" turnings on the spokes. Some slight construction differences exist among the group, but a characteristic look dominates.[5]

A generation earlier Abraham Overholt was working in Bucks County, Pennsylvania, and he left a saxony with the mark "AOH" picked into the end grain of the table along with the date, "1790."[6] He left account books from 1790-1833. The wheel and the marks are very interesting for they represent another style entirely. The noteworthy features are: 1) the "flat hat" finials on the maidens, 2) the turnings on the wheel spokes, 3) the wheel post supports which go all the way down to the legs, and 4) the use of a punch to outline a field on either side of the tension screw into which the initials and dates are picked.

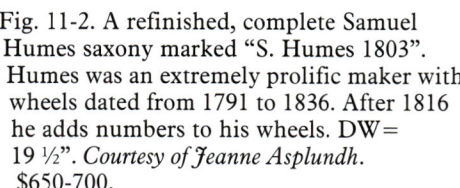

Fig. 11-2. A refinished, complete Samuel Humes saxony marked "S. Humes 1803". Humes was an extremely prolific maker with wheels dated from 1791 to 1836. After 1816 he adds numbers to his wheels. DW= 19 ½". *Courtesy of Jeanne Asplundh.* $650-700.

The first three features can be found on a group of wheels coming out of northern Bucks County which have the stamped marks "D. Reiner," "S. Reiner," and "I. Sellers" (Fig. 11-3). "A. Todd" was in Philadelphia or Bucks County and belongs in this group (Fig. 11-4). "I. Homsher" belongs in the group, but he made outlandishly large bobbin and flyer units which suggest he was supplying wool spinners with saxonies (Figs. 11-5 and 11-6).

Fig. 11-3. An "S. Reiner" wheel with an unusual pine-like, light brown finish. Very occasionally wheels with this stain appear in this region. Similar wheels marked "D. Reiner" are also found. Not being documented Lancaster County piece, these wheels draw a lesser price. DW= 19 ½". *Courtesy of Michael Taylor*. $400-450.

Fig. 11-4. A marked "A. Todd" saxony in wonderful condition. The little "flat hat" finials on the maidens along with the wheel support posts which rest on the legs are characteristics of one group of wheelmakers who resided in what is now the Philadelphia area. Abraham Overholt is the best documented of this group. DW= 19 ½". *Courtesy of Jeanne Asplundh*. $500-550.

Fig. 11-5. Here is a wheel marked "I. Homsher", which has an outrageously large bobbin and flyer unit. When these first appeared, people assumed they had been converted, but there are now enough known examples to suggest that Homsher was making these with distaffs and with these large units commonly associated with wool spinning. Not all of these are signed. DW= 18 ½". *Courtesy of Jeanne Asplundh.* $500-550.

Fig. 11-6. A close up of the surprisingly large bobbin and flyer unit on the previous wheel. Like many wheels from this area, it has a little reeling pin stuck in a hole right behind the mother-of-all. *Courtesy of Jeanne Asplundh.*

Another group, associated with Berks County, produced very colorful saxonies. They used stripes in combinations of black, red, orange and yellow. One group also had some unusual features, in addition, to the use of color. "J. Fox," "D. Kunkel," "D.S." and "H.S." worked in color and "J. Jacob" did not (Figs. 11-7 and 11-9). The features of their wheels that are distinctive are: 1) the turnings on the spokes, 2) the wheel support posts fit into a slot cut out of the table and are nailed in from the end, and 3) the drive wheels are made of four roughly equal pieces which are not fitted together radially giving the appearance of a jigsaw puzzle. "D. Kunkel" was probably Daniel Kunkel (1798-1863) of Berks County. An example of his wool wheel is shown in Fig. 2-8. A closely related wheel is marked like Abraham Overholt's with "W+M 1822 No. 28" (Fig. 4-2).[7] Other styles of saxonies in the Berks County region used color. "B. Creen," once thought to be a New Jersey maker, was probably a Berks County maker. He freely borrowed from several traditions, but the use of color, and one of the finial turnings he employed are characteristic of Berks County. "D.S." and "H.S." marked their wheels like "AOH" and "W+M."

Fig. 11-7. This wheel marked "J. Jacob" has all the idiosyncratic construction features of the "I. Fox" group, but the wheel never had any bands of color. A Jacob Jacobs is listed as an apprentice turner in Baltimore in 1802. It is likely the same person with Jacobs having moved into the Berks County, Pennsylvania area. Instead of a distaff this wheel has a Lancaster County tow fork. DW=19 ½". *Courtesy of Michael Taylor.* $500-550.

Fig. 11-8. A close up showing the odd way in which the wheel support post is cut and nailed into the table. It is one of the distinctive features of this group of makers. They also make their drive wheels of four roughly equal pieces but with uneven cuts. *Courtesy of Michael Taylor.*

The best known of the Berks County group is Jacob Fox (Fuchs). He is a documented maker of Windsor chairs. Born in 1788, he was clearly working in the first half of the 19th century. Initially he used a stamp "IA: Fox." From 1839 on he used a different stamp "J. Fox." We know this because the second stamp still exists and has the date 1839 on it. Fox worked in Tulpehocken Township in Berks County until his death in 1862. His wheels are excellent example of the Berks County group (Fig. 11-9).[8]

Only in this region of Pennsylvania did wheel makers include a "reeling pin" to protect the reeler's finger as the flax was taken off of the bobbin and onto the reel. They are found in some reels, saxonies and Irish castle wheels. The reeling pins themselves are usually missing, but the small hole in the table above and in front of the mother-of-all is a telltale sign that one was once there. It seems to have been the fashion from the 1820s on as Humes didn't have one, but both Danner and Killian incorporated it into their saxonies. Danner used the extra hole in the Irish castle crosspiece to hold one too.

Fig. 11-9. "J. Fox" is the mark of a turner who anglicized his name from Fuchs to Fox. Jacob Fox (1808-1862) worked in Tulpehocken Township in Berks County, Pennsylvania. Prior to 1839 he used a stamp with "IA Fox", so this wheel is pos- 1839. This superb example has the wonderful color bands associated with the Berks County wheels. DW=19". *Courtesy of Jeanne Asplundh.* $700-750.

In two cases the account books of makers who signed wheels have been preserved. The story of eastern Pennsylvania saxonies is hardly over as the "I H 1827" (Fig. 11-10) and the "B. Creen" wheels (Fig. 11-11) make it clear that there is much sleuthing remaining to be done. Creen uses two very different finials for his maidens, and some of his wheels have the colorful bands of red and black and some do not. The colorful bands look original to the piece. J. Fox's saxonies all seem to have the bands. D. Kunkel's wool wheel shows no evidence of color and is in "barn found" condition, but his saxony is vibrant. The best guess, of course, is that they were responding to market conditions and varied their output.

---

[1] James D. McMahon, "Daniel Danner, Woodturner of Manheim, Lancaster County: An Early Nineteenth-Century Rural Craftsman in Central Pennsylvania." Master's Thesis, The Pennsylvania State University at Harrisburg, 1992. McMahon is not an expert on textile equipment, but his work on Danner and his comparisons with Abraham Overholt's career are instructive. The account books of Danner's at The Hershey Museum make quite interesting reading, especially the account book which begins in 1835. Bill Leinbach, the noted spinning wheel collector, is a special fan of Danner's and has looked at much of his surviving work. William A. Leinbach, "Daniel Danner: The Man Behind the Wheel," *The Spinning Wheel Sleuth* 4 (January 1994): 6-7.

[2] See either the color photo in Peter S. Seibert's article, "Decorated chairs of the lower Susquehanna River valley," *The Magazine Antiques* (May 2001): 783 for photos of the Danner saxony, the 1791 Humes saxony and the 1841 Danner Irish castle wheel in the collection of The Heritage Society Museum of Lancaster County.

[3] McMahon, 57.

[4] Humes probated inventory at death is on file at The Lancaster Historical Society.

[5] Michael B. Taylor's "Humes, Danner, and Killian Flax Wheels," *The Spinning Wheel Sleuth* 35 (January 2002): 2-3. "J. Killian" is the mark of either Jacob Killian (1761-1828) or his son, John Philip Killian (1804-1885), both residing in Lancaster, Pennsylvania.. The later is listed as a turner in tax and census records in the 1838 and later. The existence of a virtually identical "P. Killian" saxony raises some interesting questions since there is no P. Killian other than John Philip Killian in the family. However, it is known that later in life Jacob Killian went by Abraham, his middle name, and it is suggested that at that time going by one's middle name was popular among the Pennsylvania Dutch. So "P. Killian" is almost certainly John Philip Killian.

[6] Alan G. Keyser, Larry M. Neff and Frederick S. Weiser, trans. and eds. *The Accounts of Two Pennsylvania German Furniture Makers: Abraham Overholt, Bucks County, 1790-1833 and Peter Ranck, Lebanon County, 1794-1817.* (Breinigsville, Pennsylvania: The Pennsylvania German Society, 1978) shows photos of the AOH wheel at the Mercer Museum as well as translates the accounts of Overholt.

[7] Margaret B. Schiffer, *Furniture and its makers of Chester County, Pennsylvania,* (Philadelphia: University of Pennsylvania Press, 1966). Since the mark is picked into the field made by punched flowers, the "+" may be a "J." William J. Major is a known Chester Springs, Chester County, Pennsylvania maker, who advertised spinning wheels in 1829.

[8] Nancy G. Evans, *American Windsor Furniture: Specialized Forms,* (New York: Hudson Hills Press, 1997):220-221. Unfortunately, the Fox wheel shown on p. 221 has a "married" wheel. As mentioned earlier, even museums can end up with compromised wheels. Nancy G. Evans, *American Windsor Chairs,* (New York: Hudson Hills Press, 1996): 127.

Fig. 11-10. This wheel, marked "I H 1827", has brilliant bands of color, but has some differences from the J. Fox group. It is certainly from the Philadelphia area. DW=20". *Courtesy of Jeanne Asplundh.* $600-650.

Fig. 11-11. "B. Creen" marked wool wheels and saxonies pop with and without the color bands and with two different finials. The color and one of the finial styles strongly suggest a Berks County origin. The wheel post support goes directly into the table far differently than the Fox group does. The great color and condition of such wheels increases their values dramatically. DW=19". *Courtesy of Michael Taylor*. $650-700.

Chapter 12
Shaker Wheels

Shaker spinning wheels look very much like many other New England wheels of the early 19th century, or perhaps it should be said that New England wheels of the 19th century look like Shaker wheels. Several Shaker communities made thousands of wheels and sold them throughout New England for almost 100 years, from the 1790s until at least the 1870s. The beautiful simple style associated with Shaker furniture came to be the dominant type of wool and flax wheel in New England. Since many wheels are unmarked, it is important to learn how to recognize the wheels from the various Shaker communities whether marked or unmarked. (See Table 1, at the end of this chapter, for a summary of marks and the Shaker communities with which they go.)

Finding a Shaker spinning wheel is not hard, and except at Shaker auctions they do not bring premium prices in spite of their beauty and ease of identification. We would not settle for a Shaker wheel except if it is marked, relatively complete, and with a good finish because they come on the market regularly. Flax wheels from Canterbury have sixteen spokes and are especially attractive. Flax wheels from Alfred and Sabbathday Lake are similar, but the finials and wheel posts are slightly different. Both frequently use dramatic quarter-sawn oak in tables and wheel rims as well tiger-striped maple legs and spokes. On Alfred flax wheels there is a noticeable thickening of the diagonal piece of the treadle assembly, so that the left foot has an easy place to rest. It can be seen on Sabbathday Lake wheels, but it is less obvious. Anyone interested in buying a Shaker wheel would do well to visit the Shaker museums at Old Chatham, New York, Hancock, Massachusetts, Pleasant Hill, Kentucky, Canterbury, New Hampshire, and Sabbathday Lake, Maine to see examples. The American Textile History Museum also has a number on display. One exceptional example is the "D.M." wool wheel at The Shaker Museum at Old Chatham with an extraordinary accelerated bat's head where one would normally find a Minor's head.

The first community, which was gathered into "gospel order," was New Lebanon, New York, in 1787.[1] David Meacham was the first trustee there. Trustees were designated to do business with the outside world, and their initials, not the maker's, appear on the wheels. "D.M." was stamped into the end grain at the high end of the table on wool wheels (Fig. 12-1). The community probably made and sold flax wheels (saxonies) as well, but to date none with the appropriate incised "D.M." stamp has surfaced. Flax wheels with a raised "D.M." stamp (typically repeated) have been sold as Shaker, but the basis for the attribution is weak. Other Shaker communities used the same stamp on both wool wheels and flax wheels. No "D.M." flax wheels have a solid provenance, whereas some "D.M." wool wheels have excellent histories of coming from Shaker communities. New Lebanon wool wheels are easily identified by their unusual tension device. Instead of the threaded rod passing through the spindle post, it pushes against the bottom of the spindle post which is supported by two short posts (Fig. 2-15). Henry DeWitt of New Lebanon was making spinning wheels into the 1840s.[2] He may be responsible for the odd little accelerated spindle heads, which the Shakers made rather than supply Minor's heads (Fig. 12-2).

Fig. 12-1. A wool wheel with the mark "D.M." for David Meacham, trustee of the New Lebanon, New York community. The tension device is unusual and does not appear on other Shaker wool wheels (Fig. 2-15). Not all Shaker wheels are marked. The accelerated head is a reproduction of one found on a "D.M." wheel at the Shaker Museum, Old Chatham, New York. "D.M." wheels are quite scarce. DW=43 ½". *Courtesy of David Pennington.* $650-750.

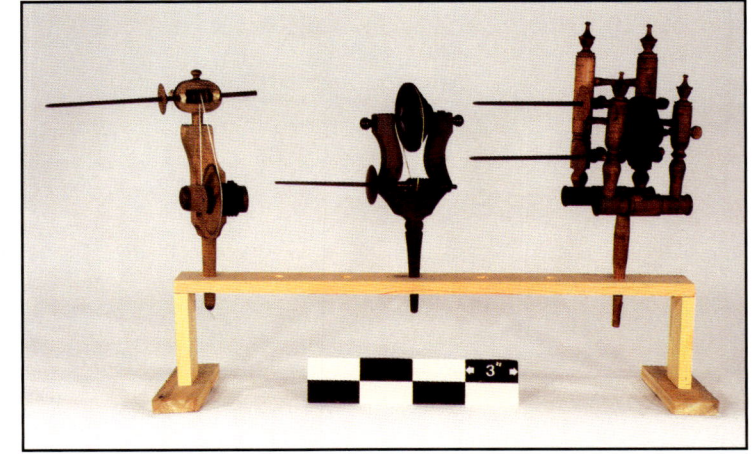

Fig. 12-2. The one on the left is the reproduction of a Shaker's Minor's head with a little removable bat's head element. *Courtesy of David Pennington.* $200-250 for an original Shaker example.

Not all eastern Shaker communities made and sold wheels. Indirect evidence suggests that the Hancock, Massachusetts and Enfield, Connecticut Shaker communities obtained their wheels from New Lebanon. The big producing communities were the Maine and New Hampshire ones. In New Hampshire it appears that Canterbury started making and selling wheels at an early date. Zadock Wright was a trustee from 1793-1807 and wool and flax wheels with his mark, "Z.W.," are known (Fig. 12-3). Since many New England Shaker wheels are unmarked as well as marked, we suspect that they marked the ones they placed with retailers in the "world," which we know they did from correspondence in the archives of The Shaker Library at Sabbathday Lake, Maine. Early trustees, such as Wright had their own mark, and when they were replaced, a new set of initials were made. So, Wright's successor as trustee in 1807, Francis Winkley, had a stamp "F.W.," which can be found on wool and flax wheels (Fig. 12-4). Winkley was replaced in 1828, but when he left his position, the Canterbury Shakers continued to use the stamp as their mark. As we shall see, the same thing happened at the Alfred, Maine community, another major producer of wheels.

Fig. 12-3. A Shaker saxony marked "Z.W." for Zadock Wright, the Canterbury, New Hampshire trustee from 1793-1807. The basic style of Canterbury flax wheel and wool wheels remained unchanged although the turnings did get a little leaner over time. DW=20". *Courtesy of David Pennington.* $600-650.

Fig. 12-4. Canterbury Shaker wool wheel marked "F.W." for Francis Winkley who became a trustee or deacon in 1807. "Z.W." wool wheels are very similar. The hoops for the rims are one piece rather than two as is the case with Alfred, Maine Shaker wheels. DW=45". *Courtesy of David Pennington.* $400-450.

Josiah Edgerly was also a Canterbury trustee associated with the wheel-making business. He was a trustee from 1811-1824. His flax wheels are signed "J. Edgerly, Canterbury," and his flax wheels are uncommon (Fig. 12-5). Francis Winkley was quite sick from 1811-1813, and it is possible that Edgerly was involved in the business only in those years. Wheel-making at Canterbury goes back to 1794. Elder Henry Blinn, the noted Shaker leader and historian of the mid-19th century, wrote that wheel manufacture at Enfield, New Hampshire, ran from 1803-1849.[3] Nathaniel Draper, Enfield's trustee in 1803 was a close friend of Winkley's, and visited Canterbury in 1801 and 1803. Wool wheels with the "N.D." stamp (Fig. 2-31) are indistinguishable from those of "F.W." Since many wool wheels with their distinctive features have no mark, we aren't sure which New Hampshire community the unmarked ones are from. Other marks associated with the New Hampshire communities are "S.M." and "B.L." The "S.M." wool wheel is probably associated with Stephan Merrill of the North Family at Canterbury, and the "B.L." may have been a presentation piece for Bennett Libbey as it has a raised mark and has been kept at the Canterbury community throughout its history.

Fig.12-5. A Canterbury Shaker flax wheel marked "J. Edgerly, Canterbury". Edgerly replaced Francis Winkley as trustee from 1811-1813 when the latter was sick. The location identification suggests they were being sold at some distance from this community or that they wanted to distinguish them from those of Enfield, New Hampshire, Alfred and Sabbathday Lake, all of which were also making wheels at this time. DW=21". *Courtesy of David Pennington.* $650-700.

The Alfred, Maine community was another major producer of spinning wheels, and as we shall see, it was a direct competitor of Canterbury. Ralph Esposito in his master's thesis, *The Development of the Wool Spinning Wheel in the United States,* studied wheel production at Alfred and found several makers listed in daybooks, including Benjamin Bailey who produced 746 between 1835 and 1862. He did not distinguish between wool and flax wheels, but Edward Goodrich made 253 "large" wheels, 24 undesignated, and one "small" wheel from 1835-1839. So in five years he made 278 wheels while Bailey made 746 in 27 years.[4] Two marked "B.B." (dyslexic "B's") wool wheels have been seen, but they may be associated with Barnabas Briggs, trustee at Sabbathday Lake 1794-1801 since they resemble that community's style. Nathaniel Freeman was both a trustee and a wheel maker. He made 117 "large" or "woolen" wheels between 1835 and 1849. Wool wheels marked "NF AL^D" have been seen (Fig. 12-6). The best-known mark on Alfred wool wheels and flax wheels is that of "SR AL." Samuel Ring was an Alfred trustee from 1809-1814. His mark is found on flax wheels (Fig. 12-7) and two very different types of wool wheels (Figs. 12-8 and 12-9). His flax wheels followed the style of Thomas Cushman who was involved very early on in the wheel business at Alfred (Fig. 12-10).

The wool wheel differences are very illuminating when coupled with a letter in the files of the Shaker Library at Sabbathday Lake which also houses Alfred's records. Dated Aug. 2, 1823, it was addressed to Mr. Issac Bracket of Alfred:

> Sir. I am nearly out of wheels. I should like to have you bring me some as soon as you can make it convenient say 6 or 9 large and 3 or 4 small. I wish you would have ears fixed to the large wheels. All the Canterbury wheels are fix'd with them, and people find fault with yours for not having them. – some small sieves likewise I should like to purchase. Your humble servt Stephen Pearse.[5]

The "ears" referred to here are the handles on the tension screws. Early Alfred wool wheels have a tension device in which the handles move since they are attached to the spindle post. Interestingly, there are examples of "SR AL" wool wheels with the "fix'd ears." Beginning around 1809 many Shaker communities continued to use their trustee marks after the trustee retired. We know it happened with the pharmaceutical labels at New Lebanon, and it would make sense of the fact that the last trustee mark used at Canterbury was that of Francis Winkley, who retired as trustee in 1828.

Wool wheel production at Alfred seemed to be far greater than flax wheels. The only marks on Alfred flax wheels are those of "T.C." and "SR AL." Some trustees were wheel makers and some were not. They moved back and forth between Alfred and Sabbathday Lake. Thomas Cushman was a trustee at Alfred from 1801-1809 and at Sabbathday Lake from 1809-1816. Wheel production began at Alfred in 1793 when they converted the old meetinghouse to that purpose and the Anderson brothers, James, John, and William set up production.[6] William made wheels for 37 years, but no "W.A" initialed wheels have been sighted; and since he was never a trustee, there probably aren't any. John was a trustee at Alfred twice (1801-1806; 1817-1822) and Sabbathday Lake once (1806-1814). "J.A." marked wool and flax wheels of the style associated with Sabbathday Lake have been seen.

Fig. 12-6. Nathaniel Freeman of the Alfred community went his own way in making wool wheels designing his own style with a nice ball top finial. His mark "NF AL^D" follows the Alfred pattern of identifying the community. A relatively scarce mark. DW=44". *Courtesy of David Pennington.* $600-650.

Fig. 12-7. A classic Alfred, Maine, flax wheel marked "SR AL" for Samuel Ring, a trustee from 1809-1814. It is ironic that because his stamp became the standard mark at the Alfred Shaker community, Ring, who did not make wheels and was trustee for a relatively short period of time, is the one associated with Shaker spinning wheel production in Maine. The use of quarter-sawn oak and tiger maple in Alfred wheels is quite dramatic. DW=19". *Courtesy of David Pennington.* $400-450.

Fig. 12-8. The earlier Alfred wool wheels had an odd tension device (Fig. 2-16). This wheel is marked "SR AL". Notice that the leg is located behind the tension system giving the wool wheel a very elongated profile. DW=44". *Courtesy of David Pennington.* $400-450.

Fig. 12-10. Thomas Cushman appears to have been the lead trustee in the wheel business at Alfred from 1801-1809. The flax and wool wheel styles were developed during his tenure and a few with his "T.C." mark have been seen. DW=20". *Courtesy of Jeanne Asplundh.* $700-750.

Fig. 12-9. After 1823 the Alfred Shakers moved to the more standard tension device used by Canterbury, Alfred, and Sabbathday Lake. This wool wheel is also marked "SR AL". DW=43". *Courtesy of David Pennington.* $350-400.

Fig. 12-11. This wool wheel marked "J.H." is from the Sabbathday Lake, Maine community. There were a number of trustees there with these initials, so we are not sure whose they were (see text). Neither "J.H." wool wheels or flax wheels are common although Sabbathday Lake was in the wheel business from 1801 until at least 1874. DW= 44". *Courtesy of James Munsie.* $450-500.

The mark commonly associated with Sabbathday Lake wheels, wool and flax, is "J.H." (Fig. 12-11). Given that either Josiah Holmes, his two sons, James and John, or Josiah's colleague, Josiah Harding was a trustee from 1801-1856, the stamp doesn't help much with dating. A classic Sabbathday Lake flax wheel with the mark "N.M." has been seen. It is probably Nathan Merrill. A small family at Sabbathday Lake was called Poland Hill. Wool wheels like the one in Fig. 12-12 are associated with the Poland Hill family.

The Shakers in western communities also made spinning wheels. Today at the Pleasant Hill, the restored community near Lexington, Kentucky, visitors can see octagonal posted wool wheels with rotating barrel tension devices with a metal band around the top of the spindle post. One can also see several of the patent spinners known as "Pleasant Spinners" there. When Mike was being given a behind the scenes tour at Pleasant Hill in the 1970s, he spotted a Lyman Wight pendulum wheel, which as far as anybody knew had been on the site when it became a museum in 1960. The Shakers were very progressive in their adoption of technology, so perhaps they bought one. Because the Shakers were rural and communal, their views on technology are sometimes confused with the Amish.

The manufacture of spinning wheels was a very successful business for the Shakers for a long time. We know that they were still making them at Sabbathday Lake in 1874. In 1812 when a western Shaker leader was making an inventory of items left at the short-lived Shaker community at Busro, Indiana, the leader mentioned that there was "timber enough prepared for 1000 big and little spinning wheels."[7] That certainly suggests they were planning on selling throughout the Ohio River Valley. They were probably influenced by the success of the Shaker communities in the east in this line of business.

The Shakers also made other textile tools for sale to the outside world. They made literally thousands of umbrella swifts at Hancock (see Fig. 14-16) as well as clock reels at New Lebanon into the 1840s. The painted dials on these reels had the initials of "D.M.," the trustee mark used on items made for "the world." They also made "patent wheel heads" and "Pleasant Spinners" for themselves and possibly for resale.[8]

Fig. 12-12. This unmarked Shaker wool wheel is associated with the Poland Hill family at Sabbathday Lake. Note the unusual tension device! DW=45". *Courtesy of David Pennington.* $450-500.

[1] Edward D. Andrews, *The Community Industries of the Shakers: Museum Handbook Number 15*, (The State University of New York, 1933): 19, 54.
[2] Jerry V. Grant and Douglas R. Allen, *Shaker Furniture Makers*, (Hanover, N.H.: University Press of New England, 1989): 72.
[3] Michael Taylor, "Shaker Spinning Wheel Study Expands," *The Shaker Messenger*, Vol. 8, No. 2 (spring, 1986): 9. Most of the information in this chapter on Shaker wheels is based on this article from *The Shaker Messenger*. From 1974 to 1986 Brother Theodore Johnson of the Sabbathday Lake community was very gracious about forwarding information from its archives as it pertained to spinning wheels. Wendell Hess, Rob Emlen, and Emily Van Hazinga were doing likewise with their researches at Canterbury and their private holdings. Eldress Frances Carr and Brother Arnold Hadd helped Mike work with The Shaker Library holdings after the tragic death of Brother Ted. Jerry Grant and Tim Rieman, both, were wonderful about forwarding tidbits about spinning wheel and textile production as they did their respective work on Shaker furniture in various archives. All of us would acknowledge that Sister R. Mildred Barker of the Sabbathday Lake Shaker community was our inspiration for careful research into Shaker history.
[4] Ralph J. Esposito, "The Development of the Wool Spinning Wheel in the United States," Master's Thesis, State University of New York College, Oneonta at Cooperstown, 1970, Appendix II.
[5] Michael Taylor, "Shaker Spinning Wheel Study Expands," 26
[6] Sister R. Mildred Barker, "A History of 'Holy Land'- Alfred Maine (Part I)," *The Shaker Quarterly*, (fall, 1963): 92.
[7] Mary Lou Conlin, "The Lost Land of Busro," *The Shaker Quarterly*, (summer, 1963): 54-55.
[8] Grant and Evans, 72.

Table 1
Marks on Shaker Spinning Wheels

Mark	Trustee	Shaker Community	Dates of Use
D.M.	David Meacham	New Lebanon, NY	1790-?
Z.W.	Zadock Wright	Canterbury, NH	1793-1807
F.W.	Francis Winkley	Canterbury, NH	1807-?
J. Edgerly, Canterbury	Josiah Edgerly	Canterbury, NH	1811-1813
S.M.	Stephan Merrill	Canterbury, NH	1825-?
B.L.	Bennett Libbey	Canterbury, NH	?
N.D.	Nathaniel Draper	Enfield, NH	1803-1849
B.B.	Barnabas Briggs	Sabbathday Lake, ME	1794-1801
N.M.	Nathaniel Merrill	Sabbathday Lake, ME	1796-1806
J.H.	Josiah Holmes	Sabbathday Lake, ME	1801-?
	James Holmes		
	John Holmes		
	Josiah Harding		
J.A.	John Anderson	Sabbathday Lake, ME	1806-1814
T.C.	Thomas Cushman	Alfred, ME	1801-1809
SR AL	Samuel Ring	Alfred, ME	1809-?
NF AL^D	Nathaniel Freeman	Alfred, ME	1822-1862

Chapter 13

Fancy European Wheels

Some extraordinary flax wheels were made for and used by upper class and wealthy women in Europe. Some spinning wheel makers embellished existing forms through the use of paint, brass, inlays, and ivory or bone turned elements. In addition, some made small toy-like examples. Others, however, went one step further and made enhanced technology the focus of their efforts.

First, we will look at traditional spinning wheel styles with expensive embellishments. Then we will turn to the more exotic.

Fig. 13-1. This example is thought to be from the Balkans. The distaff is in two sections. DW=10". *Courtesy of Jeanne Asplundh*. $400-450.

The small European saxony in Fig. 13-1 is carefully painted. The one in Fig. 13-2 has bone and ivory inlay and a row of wooden bells hanging from the rear of the table. Either of these wheels would have made a fine wedding present or dressed up a parlor nicely. They are smaller than their American cousins (see Fig. 3-1). Fig. 13-3 shows a small saxony with bone buttons and small wooden acorns hanging from the distaff assembly. A small painting of the Madonna with the date 1848 is inset into the table.

Fig. 13-2. This saxony has bone and ivory inlay. There is a big hole in the table where something, perhaps a painting (see Fig. 13-3) although many think they were for water cups for wetting fingers during flax spinning. DW=12". *Courtesy of Jeanne Asplundh.* $450-500.

Fig. 13-3. With its little wooden acorns hanging from its distaff assembly and the inset painting of the Madonna, this little saxony dated 1848 is a very nice example of how a basic wheel can be embellished. DW=12". *Courtesy of David Pennington.* $450-500.

Small vertical wheels with bobbin and flyers above the drive wheel seem to have been quite popular in Europe. We call ones with this configuration "parlor wheels" although others use this term for small fancy wheels of all types. Some have extremely long distaffs made up of two sections (Figs. 13-4 and 13-5). The tall distaffs were used for very long fibers indeed. The wheel in Fig. 13-6 has a matching free-standing, pole distaff. Such true pairs are quite desirable. The parlor wheel in Fig. 13-7 is ebonized with bone detailing and has a steel and brass flyer. It came with its original hook and a water cup. The wheel in Fig. 13-8 carries it all to such an extreme of embellishment that the wheel was not meant for use. Such "extras" increase the value of any such wheel. Even production models could be dressed up as the double-flyer parlor wheel example in Fig. 13-9 shows.

Fig. 13-4. This small vertical flax wheel with the drive wheel below the bobbin and flyer unit is quite common throughout Europe. We refer to them as "parlor wheels". This one has a very tall distaff, which is in two sections and can be taken apart for travel (see Fig. 13-5). DW=12". *Courtesy of Jeanne Asplundh.* $900-1000.

Fig. 13-5. The same wheel shown in Fig. 13-4, but with the top piece of the distaff disassembled to facilitate travel. *Courtesy of Jeanne Asplundh.*

Fig. 13-6. A very nice European parlor wheel with a matching pole distaff. The carved details match exactly. Of course, collectors value such matching pairs highly! DW=10 ½". *Courtesy of Jeanne Asplundh*. $2000-2200.

Fig. 13-7. A beautiful ebonized parlor wheel with bone details and the original water cup and hook used to pull the thread through the orifice. Completeness of this sort is extremely rare. DW=14". *Courtesy of Jeanne Asplundh*. $1300-1400.

Fig. 13-8. Here we go right over the top. The carved detail on some of the bone and ivory elements is extraordinary. The little ivory piece set into the treadle is unlikely to have ever seen a foot, nor was it intended to be so used. The wheel is fully functional. DW=11 ½". *Courtesy of Jeanne Asplundh*. $1800-2000.

Fig. 13-9. This double-flyer parlor wheel is ostensibly for increased production efficiency, but its delicate finial turnings, scroll work on the table, and ebonized tension screws suggest a place in the parlor, not the work room. Several of this type are in American collections, and we suspect that this version was fairly common in Holland or Germany. *Courtesy of David Pennington*. $400-450.

Miniature or toy wheels were likely for children or what-not shelves. We show four charming examples (Figs. 13-10, 13-11, 13-12, and 13-13). Not surprisingly, miniatures are copies of working wheels from the surrounding culture. At the extreme end of the decorative genre are the carved ivory and bone examples in Fig. 13-14.

Fig. 13-10. This, almost miniature, little wheel is based on a style commonly found in the Alps. We have always referred to wheels of this style as Swiss wheels. They were imported by the container load in the 1970's. This small one, however, is a quite a find as they commonly were much larger. DW=9". *Courtesy of Jeanne Asplundh.* $300-350.

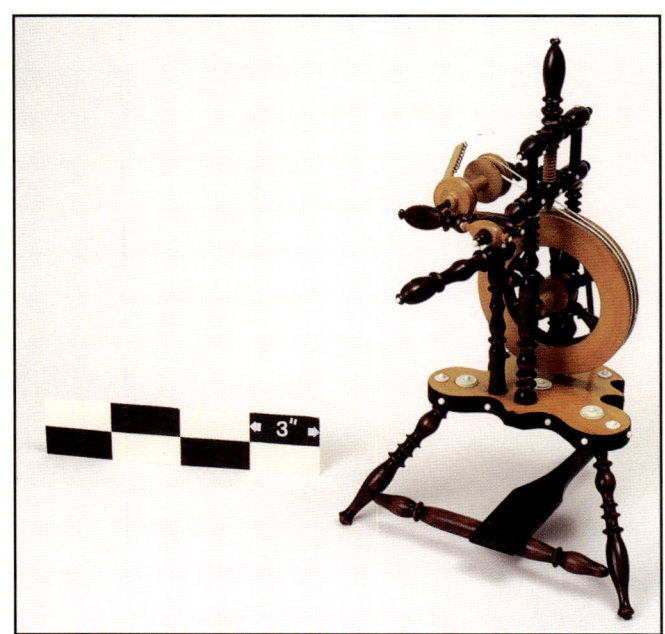

Fig. 13-11. A wonderful and very small parlor wheel with dramatic yellow paint. Sometimes these small wheels do not have working bobbin and flyer units, but this one does although it is unlikely anyone ever spun on it. DW=6 ½". *Courtesy of Jeanne Asplundh.* $650-700.

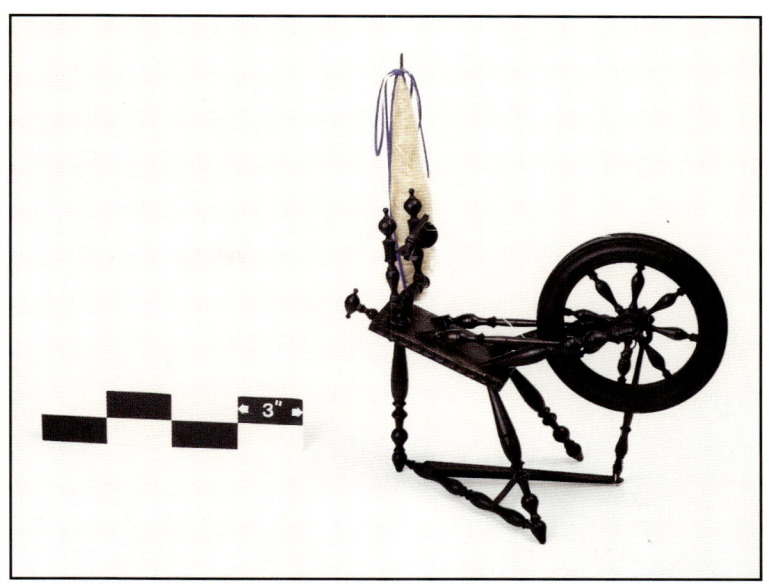

Fig. 13-12. A French style of saxony wheel, but clearly a toy as the flyer has no orifice. It is beautifully made. DW= 8". *Courtesy of Jeanne Asplundh.* $350-400.

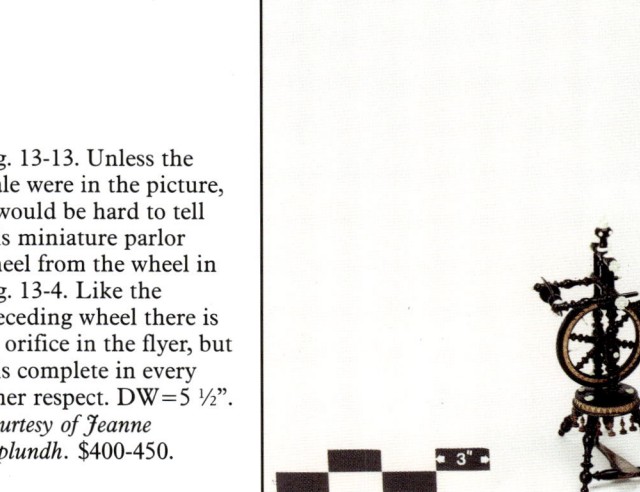

Fig. 13-13. Unless the scale were in the picture, it would be hard to tell this miniature parlor wheel from the wheel in Fig. 13-4. Like the preceding wheel there is no orifice in the flyer, but it is complete in every other respect. DW=5 ½". *Courtesy of Jeanne Asplundh.* $400-450.

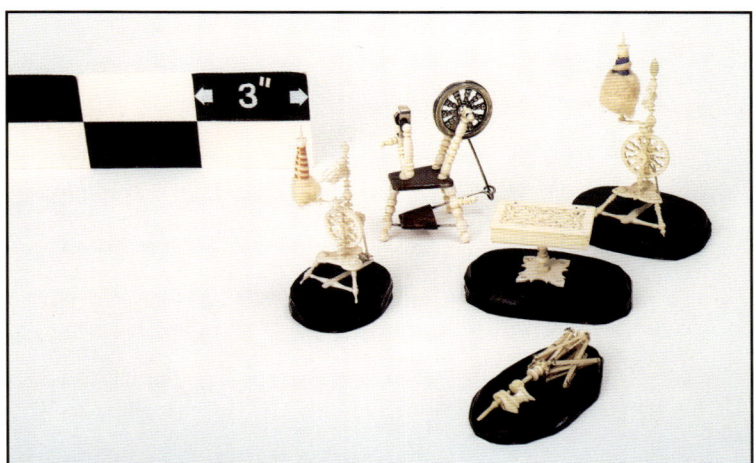

Fig. 13-14. Definitely for the "what-not" shelf, these bone or ivory carved miniatures of three wheels, an umbrella swift, and a sewing box are extremely collectible. *Courtesy of Jeanne Asplundh.* $600 each wheel. $850 for swift and table.

The French open framed wheels in Figs. 13-15 and 13-16 were illustrated in Diderot's *Encyclopedia*. The distaff is taken out of a hole in the frame and was tucked into the spinner's belt (girdle) while spinning. This style of wheel also came in a triangular base form and could also be embellished with bone buttons and finials for a fancier look (Fig. 13-17).

Fig. 13-16. In this picture of the preceding wheel the distaff has been removed and lies in front of the wheel. It would have been tucked into the belt, or girdle, of the spinner. Without Diderot's illustration from the 18th century, we would probably doubt such an arrangement. *Courtesy of Jeanne Asplundh.*

Fig. 13-15. This French flax wheel is very similar to one shown in Diderot's *Encyclopedia*. We used to refer to both the four legged and the three-legged versions as "magazine rack" wheels. If the drive wheel is missing, perish the thought, you can always use one for a magazine rack. It is particularly interesting because of the distaff and the way it was used. In this picture, we show it in its hole where it looks a little ungainly and perhaps a later addition, but it is correct. The distaff rests there when not in use. DW=16". *Courtesy of Jeanne Asplundh.* $900-1000.

Fig. 13-17. This example is very similar in style to the preceding wheel, but it has a triangular base rather than a rectangular base. This one is smaller and fancier with ivory buttons for decorations, but it does not have the Diderot association. DW=12 ½". *Courtesy of David Pennington.* $350-400.

Small ornate wheels for table top use sometimes had fancy half-spokes, and this one (Fig. 13-18) also has a brass cup and a metal flyer.

Fig. 13-18. Table top wheels like this one had obvious advantages for ladies who spun as a gentile activity, either alone or in a small group. It takes up little room and is very transportable. This one has a simple elegance that is accentuated by the exquisite brass water cup for wetting the spinner's fingers for flax spinning. Some of these have weights in the table to keep them from moving during spinning. Clamps would detract from the aesthetic and perhaps mar the fine wood of the table on which it rested. DW=10". *Courtesy of Jeanne Asplundh.* $2100-2300.

Fig. 13-19. A classic German example of the "boudoir" or carriage wheel. This one has a flyer drag arrangement by which flyer speed can be slowed by an adjustable brake. The English examples typically have a double drive cord tension system. The rims usually have a lead or pewter rim insert to increase the drive wheel's momentum. These wheels are very dainty. Some have little drawers under the table. They were surprisingly popular in England as Baines mentions in her book, and a number of them have been imported to the United States. DW= 9". *Courtesy of Jeanne Asplundh.* $1200-1300.

Fig. 13-20. The term "carriage wheel" is sometimes used in place of boudoir wheel because they disassemble readily as this photo of the preceding wheel shows. *Courtesy of Jeanne Asplundh.*

By the 1780s spinning had become a fashionable activity for "ladies."[1] In York, England there are some well documented wheel makers, and fortunately a few examples of their work have survived. "Boudoir" wheels appear to have originated in Germany and then to have become popular in England (Fig. 13-19). John Jameson made this type of wheel in York as early as 1780 and at one point indicated that he was making wheels of the type used in Germany. Since six examples of this type of wheel are in the Bayerisches Nationalmuseum in Munich, the connection seems probable. Their ornate turnings and spindly frames suggest a very refined household and careful use. Some are convertible for transportation or perhaps use in a carriage (Fig. 13-20). The table may be removed from the legs, and a handle in front allows the spinner to turn the wheel by hand rather than to treadle the wheel. The one signed by Jameson has a doubled band rather than a single drive cord with a flyer drag, which is seen on German examples. The drive wheels often have a pewter rim or a metal insert in the wooden rim to give them sufficient weight for easy spinning.[2]

Other similar forms without the elaborate gallery rails are no less exquisite. The one in Figs. 13-21 and 13-22 is wonderful. It is made of exotic woods and has a plated "Dutch gold," a bronze alloy, flyer.

Fig. 13-21. At first glance this carriage wheel looks less exotic than the previous one, but it is quite extraordinary. It has hundreds of inset hand cut steel beads, which were quite fashionable. The drive wheel rim is of hornbeam and the rest of the wheel is imported South American tulip wood. The use of such imported woods was fashionable in the 1780's when these wheels were in vogue. DW=12 ½". *Courtesy of Jeanne Asplundh.* $8300-8500.

Below:
Fig. 13-22. A close up of the wheel in Fig. 13-21 showing the cut steel beads and the metal flyer which is made of "Dutch gold", a brass alloy. This wheel disassembles like the wheel in Fig. 13-20 and hence its designation as a carriage wheel. It is possible, of course, that over the centuries the stand and the table top wheel could be separated, so it takes a sharp eye to know if there once was a stand. *Courtesy of Jeanne Asplundh.*

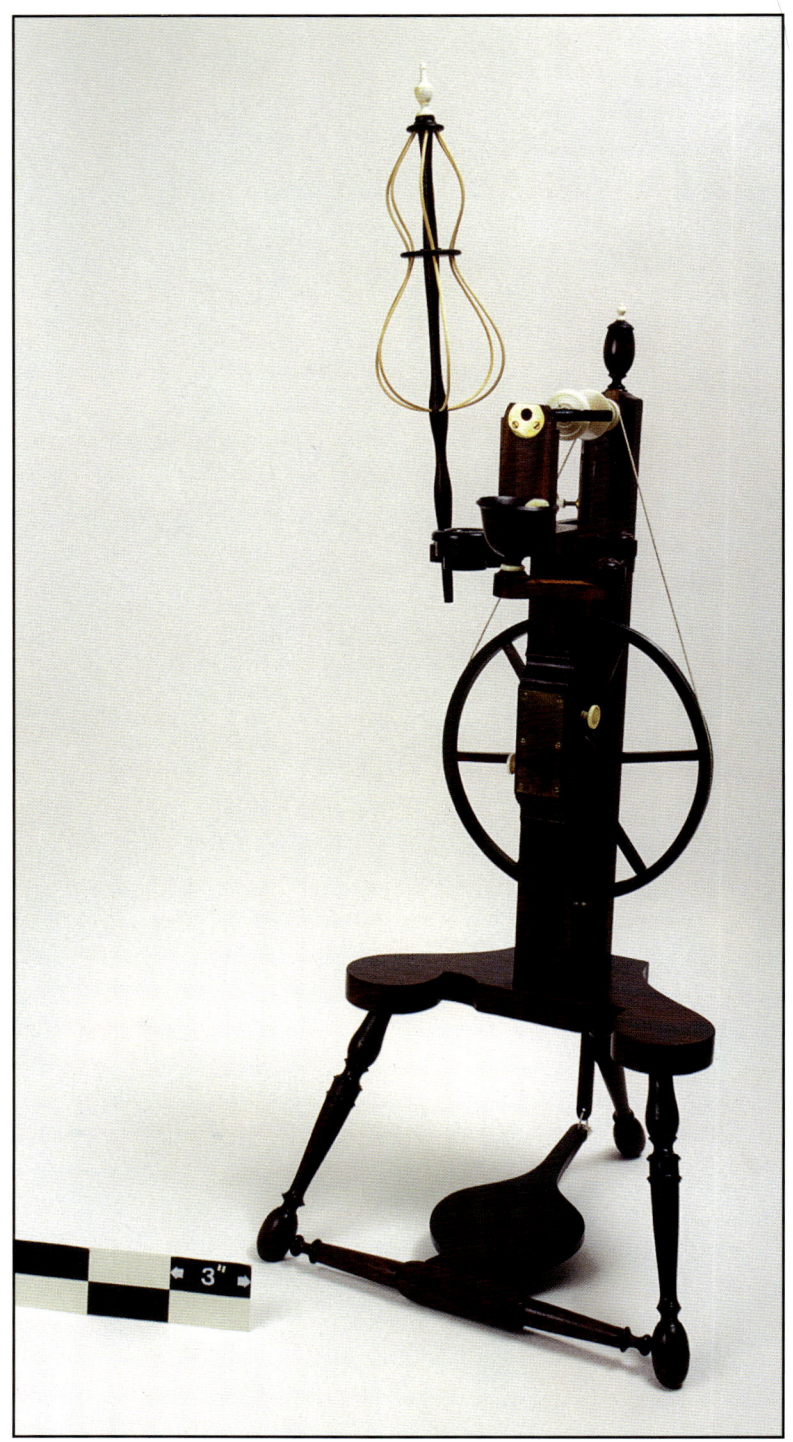

Jameson referred to himself as a "toy" maker, and others in York were likewise makers of fancy wheels and other toys. Joseph Doughty began working in his father's fishing tackle shop and developed the metal and woodworking skills necessary to produce one of the most advanced spinning wheels ever made (Figs. 13-23 and 13-24). John Antis of Fulneck, near Leeds, had received a prize for inventing a mechanism so that the spinner did not have to move the thread from one hook on the flyer arm to another in the course of spinning. Joseph Doughty made and sold these wheels from about 1795 until 1801 when he died. However, his wife and his partner, a Mr. Marshall, continued the business until 1824. By then she had changed her name to Marshall, perhaps as the result of marriage.[3]

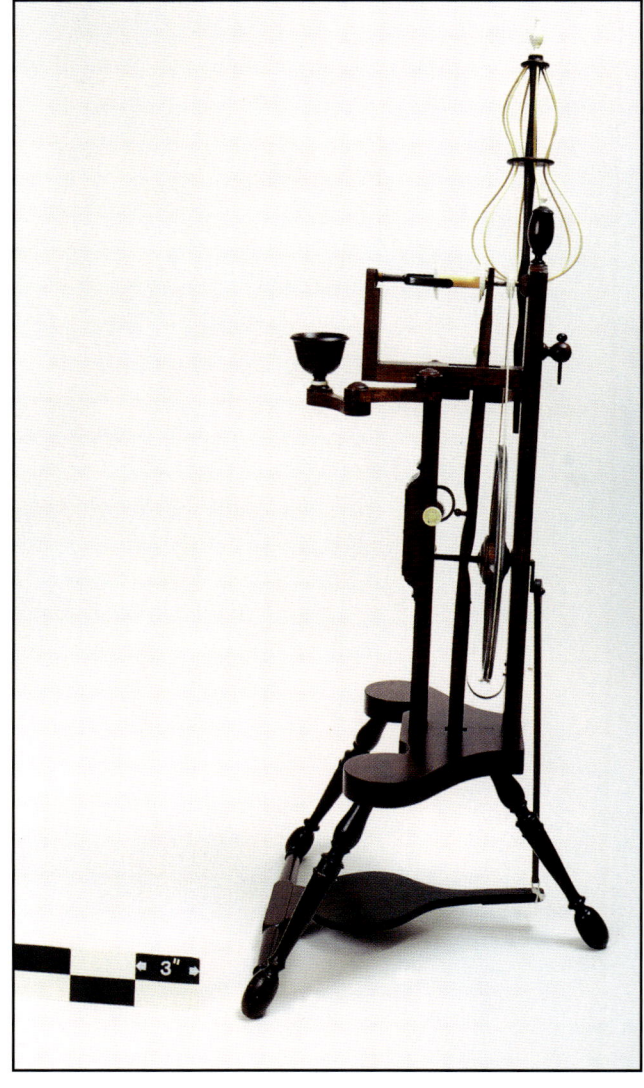

Fig. 13-23. The Doughty parlor wheels are the ultimate parlor wheels! Not only are they beautifully crafted, but they have an elliptical cam, which as it rotates, moves the bobbin back and forth. So the winding on process is fully automated, and no flyer hooks are needed. John Antis of Fulneck, near Leeds, England received an award for this mechanism and Doughty made them. The drive wheel is brass. Obviously, we are looking at one of the most desirable of all spinning wheels. DW=11 ½". *Courtesy of Jeanne Asplundh.* $9500-10,000.

Fig. 13-24. A side view of the Doughty wheel which gives a better idea of how the mechanism operates. Doughty signed his wheels, so there is no mistaking them. *Courtesy of Jeanne Asplundh.*

Fulneck was a Moravian settlement near Leeds, and John Planta was a Moravian living there, who made wheels similar to Doughty's. He also used Antis' improved mechanism, but in a horizontal arrangement which included a table. It is known that Planta was a woodworker. The Moravian group to which he belonged included clockmakers, so perhaps others in his community did the fine metalworking which requires the skills and tools of a clockworks maker. Baines suggests that Planta was probably producing these wheels only between 1798 and 1802.[4] Because a few of these wheels have a discreetly placed Planta label in the drawer, he is given the credit. These simple, but elegant, wheels are one of the highlights of this extraordinary moment in the history of the spinning wheel (Figs. 13-25, 13-26, and 13-27). Toymakers indeed!

Fig.13-25. It is said that John Planta made parlor wheels similar to Doughty, but the type of wheel shown here is the one for which he is famous. An elegant piece of furniture with an extraordinary spinning mechanism built in, it makes use of Antis's improvement. Several of these are in museums in Great Britain and Colonial Williamsburg has one, but they are much sought after. DW= 11 ¼". *Courtesy of Jeanne Asplundh.* $21,000-23,000.

Fig. 13-26. A rear view of the Planta wheel. The drive wheel is pewter. *Courtesy of Jeanne Asplundh.*

Below:
Fig. 13-27. This top down view of the Planta wheel shows clearly how the mechanism works. It also highlights the furniture aspect of the table. *Courtesy of Jeanne Asplundh.*

152

Girdle wheels in the late 18th century were made by clockmakers and used geared teeth rather than belts or cords, and the works are enclosed.[5] The one shown in Baines book and made by an English maker, James Webster, dates from the last half of the 18th century. James went into the clock-making business in 1745, and his son, Robert, gave such a wheel to the queen in 1791.[6] Some were made in Germany as well and the example shown in Figs. 13-28 and 13-29 is thought to have come from there. Jeanne Asplundh was kind enough to demonstrate hers in use (Fig. 13-30).

Several articles about these wheels have appeared over the years, and it was thought that girdle wheels were a late 18th century phenomenon until Alan Raistrick pointed out that girdle wheels go back at least to 1688.[7] Randle Holme describes a "girdle wheel" in his book, *Academy of Armory*, which he published in 1688 but started work on in 1649. It is worth quoting from the book at length because thanks to the sharp eye of Jeanne Asplundh, we can now show an example of precisely such a wheel (Figs. 13-31, 13-32, 13-33 and 13-34). It has a paper label "by Teasdale (———)Derbyshire."

Fig. 13-29. A close up of the girdle wheel. *Courtesy of Jeanne Asplundh.*

Fig. 13-28. This German girdle wheel is a frustrating wheel to picture as the mechanism is encased. The German models have a single horn to slip inside the belt or girdle, and the English examples seem to have two horns which sould be a more stable arrangement. The small metal object next to the wheel is the distaff. The wooden base or "horn" is 6 1/4", and it is 7 1/4" high. *Courtesy of Jeanne Asplundh.* $3500-4000.

Fig. 13-30. Jeanne Asplundh in period dress with her late 18th century girdle wheel. *Courtesy of Jeanne Asplundh.*

Fig. 13-31. As shown here this extraordinary wheel looks like a small 18th century table top wheel with very nice delicate turnings. However, further research revealed it to be a 17th century girdle wheel of rather large proportions compared to the late 18th century metal versions. The wheel could clearly be used on a table top as well. A paper label on the bottom of the table reads "by Teasdale (———) Derbyshire". DW=5". *Courtesy of Jeanne Asplundh.* $12,000-15,000.

This is Randle Holme's description:

> The third is the **Girdle Wheel** or the **small Wheel**; it's a wheel so little that a Gentle-woman may hang it at her Girdle or Apron string and Spin with it, though she be about.
>
> It is made and composed of **Wood, Brass,** and **Iron**: having two Wheel with Nuts on the Spindles with several other giggam bobbs pleasing to Ladies that love not to toil themselves with this sort of work: therefore may fitly be termed the **Do-Little-Wheel**, whose parts are these.
>
> The Stock to which all the other work is fixed.
> The Frame
> The Feet.
> The pillars which holds up the piece in which the Brass Wheels are.
> The greater Brass Wheel which hath Forty teeth in it, which turns.
> The lesser Brass Wheel or Nut, which hath Twenty Teeth in it, which turns.
> The small Wheel of wood.
> The Wheel string which comes from it to the Feathers.
> The Feathers, Spool, Whorve.
> The Distaff which hath a Standard, and Cross piece.
> The Handle and Axle-Tree.
> The Hooks by which it hangs to the Apron string, or Girdle.[8]

At first glance the Teasdale wheel shown appears to be a table wheel. Only after Jeanne Asplundh saw past the visual frame of reference and thought about the individual elements in it, did we see the connection between the Teasdale wheel and a girdle wheel.

Fig. 13-32. Jeanne Asplundh in 17th century period dress with her Teasdale wheel as it would have been used as a girdle wheel. Not only is she the lucky owner of this fabulous wheel, but it was her careful reading of Randle Holme's 1688 description of a girdle wheel (see the text) that revealed that this wheel was not just a table top wheel, but also a girdle wheel. During the restoration of this wheel, the wheel repairman found traces of the silk ribbon loops which would have held the hooks which fasten it to the girdle. *Courtesy of Jeanne Asplundh.*

Fig. 13-33. The Teasdale girdle wheel from another angle. This spinning wheel is of a type we have never seen an example of before, a primarily wooden girdle wheel. They were common enough for Holme to chronicle them, but few, perhaps only this one, have survived the ages. *Courtesy of Jeanne Asplundh*.

Fig. 13-34. The label on the base of the "Teasdale" wheel. So far we know nothing about him and can only assume that he was the maker. *Courtesy of Jeanne Asplundh*.

There are, however, another group of ladies wheels which are called table top wheels, and they seem to be French and from the middle of the 18th century. The 1792 French book, *Manuel du Tourneur*, describes how to make this type of wheel (Fig. 13-35 and 13-36). An exquisite one was for sale in a French antiques shop for $35,000! Less pricey ones also have been seen.[9] A piece of Meissen porcelin from around 1750 shows one on a small table next to the figure. A plain wooden example from Scotland is shown in Baines, but she also mentions one example with tortoiseshell, mother-of-pearl and ivory.[10] It is clear that with a certain class of women, spinning was partly play and hence the designation of these as toys was appropriate. They do, however, work and could be used to spin the fine linen thread used in embroidery or lace. The wheel shown in Fig. 13-37 is somewhat reminiscent of the Planta wheel in being a small side table with a spinning wheel built on top of it. The top of the table is beautifully inlaid. It is European, but we do not know which region it came from.

So far we have not come across little exquisite wheels made in America. Given the importing of fine furniture, books and scientific instruments by wealthy planters in Virginia in the 18th century, it is likely that the desire for such symbols of wealth and taste among the colonials were met by importers. By the 1840s, beautiful wheels, like the John Burley wheel Figs. 6-30, 6-31 and 6-32), were being made in America for the upper classes.

---

[1] Peter C.D. Brears, "The York Spinning Wheel Makers," *Furniture History: The Journal of the Furniture History Society*, XIV, (1978), 19.
[2] Patricia Baines, *Spinning Wheels, Spinners and Spinning*. (New York: Charles Scribner's Sons, 1977), 158-160.
[3] Ibid., 165-167.
[4] Ibid., 168-170
[5] Jeanne Asplundh, "The Search for the Elusive Girdle Wheel," *The Spinning Wheel Sleuth*, 33 (July 2001), 2-3 and Jane Lenderman-Kruse, "A Girdle Wheel by Fosters of Carlisle," *The Spinning Wheel Sleuth*, 33 (July 2001), 4-6 show enclosed gears of a similar sort. The German examples have a single horn for inserting in the girdle whereas the English examples have two horns which would appear to be more stable in use.
[6] Baines, 163.
[7] Alan Raistrick, "Extracts from the *Academy of Armory* by Randle Holme of Chester, 1688," *The Spinning Wheel Sleuth*, 20 (April 1998), 2-4.
[8] Ibid., 4.
[9] Jane Lenderman-Kruse, "Two French Tabletop Spinning Wheels," *The Spinning Wheel Sleuth*, 21 (July 1998), 6-8.
[10] Baines, 161.

Fig. 13-35. This fabulous little wheel may have come from the Paris shop of Louis Eloy Bergeron. There are eleven pages of text and illustrations of this wheel and how to make it in Bergeron's book, *Manuel du Tourneur*, published in 1798. In May 1997 there were two of these beauties for sale in Paris! Others have been seen over the years. Bergeron's shop was open for over fifty years; and since he published a detailed account of how to make one, others with similar patience and skill might have also. DW=6 ½". *Courtesy of Jeanne Asplundh.* $25,000-27,000.

Fig. 13-36. The little table top wheel shown on a table with a brass swift of the sort that might have accompanied such a wheel. *Courtesy of Jeanne Asplundh.*

Fig. 13-37. A wonderful little wheel built on an inlaid table. In this example the wheel is not detachable. It is the only such wheel we have seen, so we are unsure of its place of origin in Europe. DW=9 ¼". *Courtesy of* David *Pennington*. $2500-3000.

Chapter 14
Spinning Accessories

Spinning wheels were seldom used by themselves, as they are part of a process which begins with raising the animal or plant for the fiber and concludes with a finished textile product. Some textile historians have suggested that it took seven carders to keep one spinner supplied and about seven spinners to keep a weaver going. In her brilliant book, *The Age of Homespun*, Laurel Thatcher Ulrich, using old journals describing spinning bees and some calculations based on existing textile equipment, makes the point that for every three spinners you also need one reeler.[1] She also points out that by the early 19th century many families had access to a local carding mill, which was one of the first "masheens" of the industrial revolution and which increased household textile productivity dramatically.[2] Unlike spinning which requires some skill, reeling requires virtually none and can be quickly picked up by a young child, but it does require a reel, also known as a winder. (They are erroneously referred to as "weasels" by craftspeople and even some museum docents. "Pop goes the weasel" had nothing to do with a reel.) Once the bobbin or spindle of the wheel is full, the thread needs to be taken off. If it is to be sold for "egg money" or used in weaving, it will need to be reeled to establish length and diameter. (The diameter of evenly spun yarn can be expressed as a ratio of the weight to length or vice versa depending on the yarn count system employed.) So reels or winders are certainly one accessory that most spinners would have had with their wheels.

In the United States winders generally come in two sizes.[3] Most reels, or winders, in New England and many elsewhere are 72-75 inches (or approximately two yards) in circumference (Figs. 14-1 and 14-2). Many reels have a mechanism, which snaps or clicks after forty revolutions. That makes sense as two of the more common yarn count systems (560 and 1600) are multiples of 80 yards. In Pennsylvania winders, one often finds larger circumferences, almost always 90-94 inches (about 2 ½ yards) with snaps or clicks after 120 revolutions (Fig. 14-3). Some in Pennsylvania also used the two-yard circumference with a 150 revolutions to the click (Fig. 14-4). These winders are using a 300 yard system used to measure linen thread in Scotland and Ireland, which apparently became the common yarn count system in parts of Pennsylvania.[4] Some reels have "clocks" which show the number of revolutions since the last "click" (Fig. 14-5). A very few have multiple clocks. The slapper was sometimes shaped like a hammer. On one reel a carved head of a black man pops up, like a jack-in-the-box at the appropriate number of revolutions. Winder circumferences are almost always slightly larger than their base system requires. So a winder based on a two yard system is almost always 74-75" in circumference, a 2½ yard winder is likely to be 92" in circumference. (Yes, in the Maritime Provinces of Canada there are metric winders using two meters and 100 turns to the click!).

Fig. 14-1. Collected in Rhode Island, this little four-armed reel (74" circumference) has a very old red finish, various initials scratched in it, old hand-forged nails, and an old style of base. Not that elegant, but an ancient survivor. Most winders are in the $100-200 range unless something is special. *Courtesy of Michael Taylor*. $250-300.

Fig. 14-2. "Mutt and Jeff". Two very nice executions of the four-armed reel associated with New England and areas in the Midwest settled by New Englanders. The little one on the right was probably used on a table which is quite unusual. The tall one on the left is absolutely elegant. *Courtesy of David Pennington.* Tall one $300-350. Short one $150-200.

Fig. 14-4. A wonderful reel for several reasons. It is a six-armed (74") 150 turns to the click Pennsylvania reel. Reel circumferences are almost always slightly larger than their base system requires. So a winder using based on a two yard system is almost always 74-75" in circumference, a 2½ yard winder is likely to be 92" in circumference. Some Pennsylvania winders have reeling pins as this one does. It also has a bobbin holder and an amazing paper label (see Fig. 11-5). *Courtesy of Jeanne Asplundh.* $400-450.

Fig. 14-3. With an unattractive finish and not visually stunning, this reel is clearly marked "A.R.1790" with punch designs. It is a wonderful example of an early six-armed (92" diameter) Pennsylvania reel. It has a clock dial with beautiful paper label with a nice design and 120 marked out in units of ten. The mechanism making the click at the 120 mark is a shaped wooden hammer inside the wooden box. Most clickers are just wooden slats that slap against the side of the box, so this one is special if not beautiful. *Courtesy of Jeanne Asplundh.* $300-350.

Fig.14-5. Close up of the label on Fig. 14-5. The lettering is so small that it is probably a decal. "E.P. Rose" is the name on the label. *Courtesy of Jeanne Asplundh.*

Reels, or winders, typically have four or six arms if they are of the "windmill" variety. Most New England winders are four-armed with one interesting six-armed, slanted table variation found in Connecticut (Fig. 14-6).[5] Another very striking early form of winder associated with Connecticut is three-legged "anthropomorphic" type which resembles a person with its hands over its head. The one shown in Fig. 14-7 has red paint with some gold trim and "1777" on the base. Most of Pennsylvania and Appalachia have six-armed winders regardless of circumference. Daniel Danner, the famous maker of Irish castle wheels from Manheim, Pennsylvania, made a very distinctive five-armed winder with the usual 90+ inches of circumference (Fig. 14-8).

Fig. 14-7. This style of reel looks to some people like a person with her hands over her head. Some nice early paint and the date "1777" put this reel in a special category, but still far from best in show. Again probably Connecticut. *Courtesy of Michael Taylor.* $450-500.

Fig. 14-6. A number of these six-armed, two yard, reels with slanted table are associated with Connecticut. This one is marked "DOW". *Courtesy of David Pennington.* $250-300.

Fig. 14-8. Super five-armed, 2 ½ yard, signed "Daniel Danner, Manheim, 1828" paper label reel! These were sold by themselves and occasionally with a spinning wheel. Danner reels are far harder to find than his highly desired Irish castle wheels. *Courtesy of Michael Taylor.* $600-700.

"Double dial" winders have two hands going around simultaneously. The utility of this arrangement is questioned by contemporary weavers using traditional equipment, so perhaps it is a "bell" or "whistle" (Fig. 14-9) added by the maker. Figs. 14-10 and 14-11 show the best of the best in American reels. Signed "E. Frost," dated 1836, double dialed and with a heart shaped base, it must have been the ultimate wedding or anniversary gift!

Reeling the spun fiber from a bobbin onto a winder generally requires the reeler to use a finger to guide the thread onto the appropriate place on the reel arm. At low spinning speeds flax will eventually shear off a metal hook, so imagine what it will do to a finger at high reeling speeds. So, some spinning wheel and reel makers in eastern Pennsylvania included a reeling pin in either the wheel or the reel. In a similar vein, some wool wheels were turned with a turned dowel with a head and neck on one end. If used over a long period of time, these "wheel boys," or "wheel fingers," will leave characteristic flattening on one side of the wheel spokes. The wheel boy will also show wear if used that way. (One New England dealer called them "toddy sticks" and said they were also used as small pestles in mixing toddies. She would only sell them as wheel boys if they showed wear around the neck.) These wheel boys were probably most frequently used in conjunction with the availability of wool pencil roving from a carding mill. This pencil roving could be spun quite rapidly as it required very little, if any, "drafting, i.e. pulling apart of the fibers by hand. Fig. 14-12 shows some wheel boys.

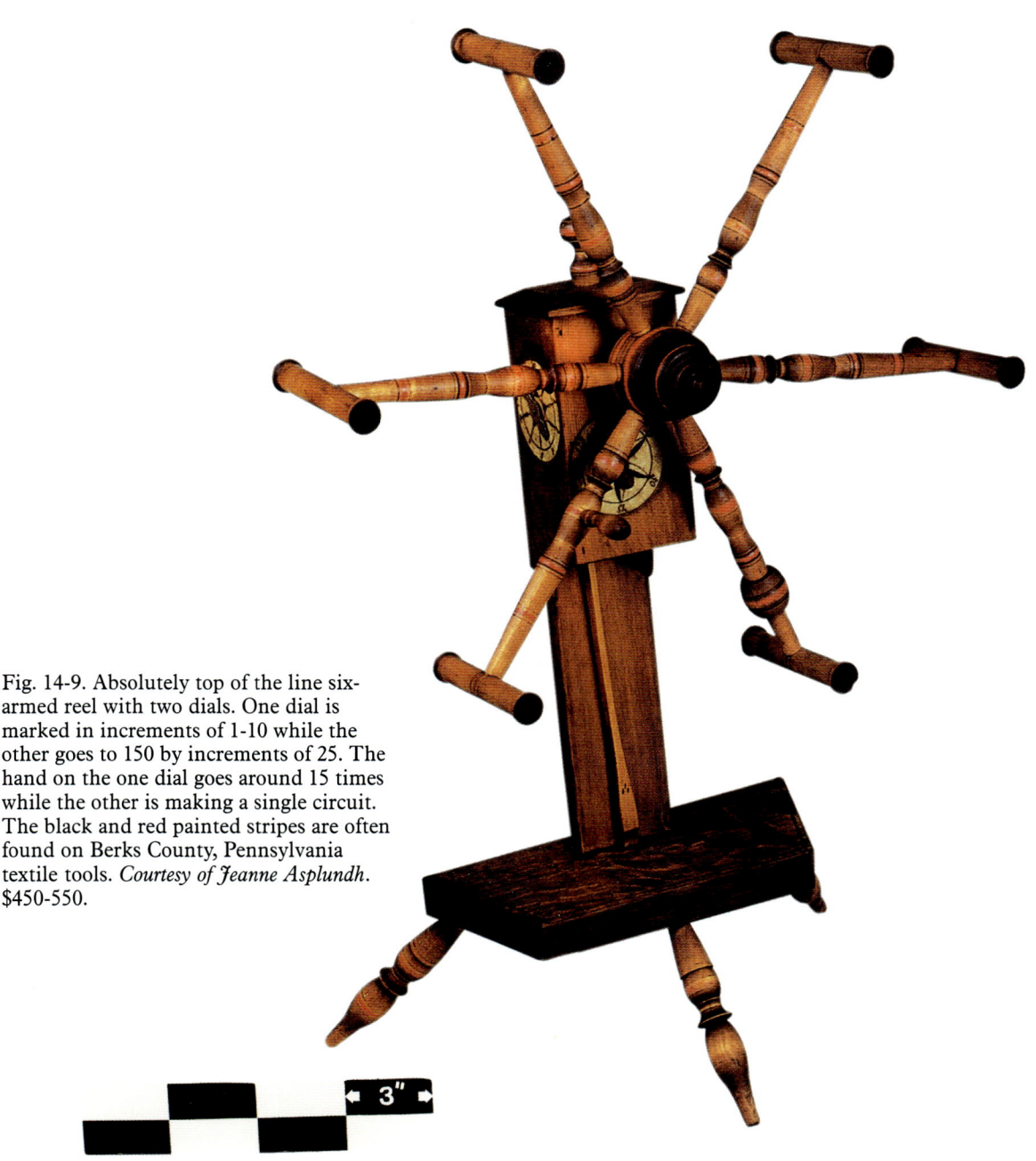

Fig. 14-9. Absolutely top of the line six-armed reel with two dials. One dial is marked in increments of 1-10 while the other goes to 150 by increments of 25. The hand on the one dial goes around 15 times while the other is making a single circuit. The black and red painted stripes are often found on Berks County, Pennsylvania textile tools. *Courtesy of Jeanne Asplundh.* $450-550.

Fig. 14-10. Signed "E. Frost", dated 1836, double dialed, and with a heart shaped base it was probably an anniversary gift. Best wooden American textile tool! Only the "John Burley" spinning wheel is a competitor. *Courtesy of James Munsie.* $1000-1100.

Fig. 14-11. Close-up of the "E. Frost" reel. *Courtesy of James Munsie.*

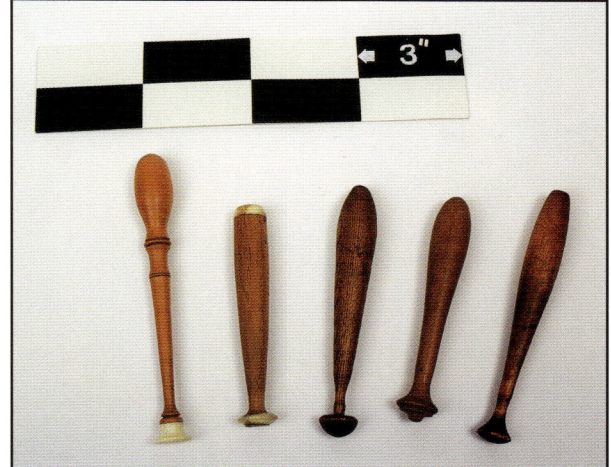

Fig. 14-12. Four American wheel boys on the right and an English boxwood example on the far left. These were used to turn a wool wheel rapidly. With the widespread availability of Minor's heads and pencil roving, wool spinners did not have to draft as they spun and could produce yarn quite rapidly. Occasionally one finds wool wheels with flattened areas on one side of the spokes which are likely to be the result of the continual use of a wheel boy. The inside of a wheel boy's neck will often show evidence of use. The same implement was used as a toddy stick, so the wear confirms its use as a wheel boy. *Courtesy of Jeanne Asplundh.* $5-10.

Fig. 14-13. Five niddy noddies, one with a skein of yarn. Once around the skein is about 74" on most American ones. Occasionally smaller ones appear and they were certainly used in 18th century Holland. Whether the small ones which occasionally appear here are imports is not clear. *Courtesy of Jeanne Asplundh.* $60-70.

Fig. 14-14. Nicely decorated niddy noddies with bracing and chip carving are rare. This one signed "M.A.H." and dated 1888 seems an anomaly with such a late date. This example might be worth more if it was not dated so late, but it is great. *Courtesy of David Pennington.* $250-300.

Another type of winder is the "niddy noddy," which is a set of three shaped pieces of wood fastened at right angles to one another. The central shaft is held in the hand, and the thread is wound around the four corners formed by the ends of the crosspieces (Fig. 14-13). When removed from the niddy noddy, a skein results. Spectacular presentation pieces sometimes appear (Fig. 14-14). At the end of the chapter is a Gallery of Accessories, which include many additional interesting examples.

A weaver or knitter with a reeled skein of yarn will wind it on bobbins to warp a loom, wind quills for shuttles, or turn it into a ball for knitting. The tool for that is called a swift. (It makes the unwinding go swiftly.) Umbrella swifts are one common form of swift. The Shakers at Hancock, Massachusetts produced literally thousands of them, most of them often in a distinctive yellow stain (Fig. 14-15). For reasons we don't understand, they produced them in, at least, five distinct sizes (Fig. 14-16).[6] As mentioned above, most skeins are either 72+ inches or 90+ inches, so it would seem that two sizes would be sufficient. Umbrella swifts in scrimshaw occasionally appear at nautical auctions, but they are quite pricey as well as very beautiful (Fig. 14-17). Of the wooden variety our favorite is a large type from Pennsylvania which has a "guitar peg" in the top section which can be turned to draw up the arms of the swift (Fig. 14-18), but umbrella swifts come in a wide variety of forms, including some all-metal varieties.

Fig. 14-15. A Hancock Shaker swift in its traditional yellow paint. This one is opened. There were thousands made from the 1840's through 1870's at the Hancock, Massachusetts Shaker community. The wooden slats are sometimes missing as the outer ones are simply tied on by string. Many are in a brown stain rather than the vibrant yellow. Completeness and vibrant color, as well as size, affect value. *Courtesy of David Pennington.* $200-225.

Fig. 14-16. Five Hancock swifts showing different sizes. Some have the cup top and some the ball top. A very few have metal bases, but the wooden base is the most common and desirable. Brother Thomas Damon listed three sizes in the 1850's, but we have found five. The four on the left are in order by decreasing size. The next to the smallest size is missing but has been collected. The best measurement of size is based on the overall height from table top to the top of the swift. The shortest one is 15 inches and the tallest one is 21 ½ inches with each one about an inch and half different from the next size. The one on the far right has a metal base. *Courtesy of Michael Taylor.* The three larger sizes all are about $200-250. The little one is $500-550.

Fig. 14-17. Scrimshaw swifts are made from whalebone and are subject to federal regulations. Their age must be verified. They are scarce but do turn up at nautical auctions. They are pricey as the scrimshaw people are very competitive. Since they were made to show off skills in carving and decorative imagination, they can be quite elaborate and few are alike. *Courtesy of Jeanne Asplundh.* $2000-2500.

Fig. 14-18. A very large umbrella swift suitable for the large skeins coming off of the 90+" Pennsylvania reels. It has a guitar key-like piece at the top which has a small cord tied to the bottom slat holder on each side. By turning the key the string winds on the key and the swift is pulled open after which the tapered key can be jammed tight and the desired position kept. Unlike the Shaker models, this umbrella swift did not have cross pieces, just one slat going diagonally from the top stay to a bottom one. A virtually identical example is the Philadelphia Museum of Art and signed "H. Kaufmann" on the clamp. *Courtesy of Michael Taylor.* $300-350.

Squirrel cage swifts look nothing like umbrella swifts in spite of their similar function. Daniel Danner called them "squirrel cage trillers" in his account book. No labeled triller of his survives although one with the appropriate turnings is shown (Fig. 14-19).[7] Pennsylvania examples are particularly impressive as they have the extra size needed to handle the larger skeins common in Pennsylvania. Some beautifully simple and plain Shaker examples can be seen in Shaker museums. Some very pedestrian examples of squirrel cage swift appeared in the mid to late 19th century as well as earlier, but some of the larger, earlier examples are stunning (Figs. 14-20 and 14-21). New England examples are not in the same league (Fig. 14-22), but a good early squirrel cage swift is always a find. Fig. 14-23 shows a squirrel cage with two sets of barrels to make handling two skeins simultaneously easier than one pair of long barrels.

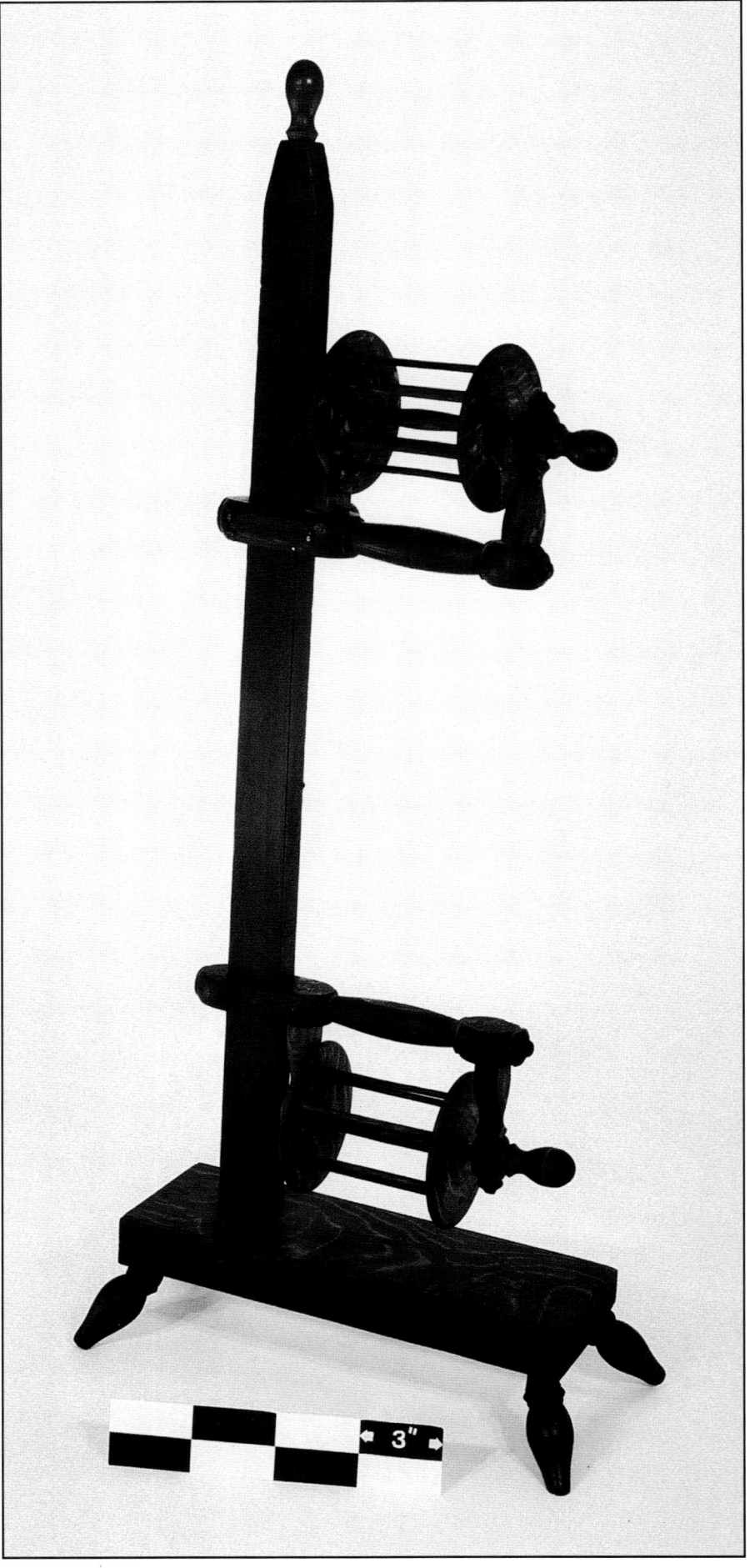

Fig. 14-19. A very nice example of a large Pennsylvania squirrel cage swift. The turnings on the horizontal pieces are very close to those used by Daniel Danner and the Landisville, Pennsylvania maker of Irish castle wheels. Danner made "squirrel cage trillers", and this example is probably one of his, but no labeled examples are known. *Courtesy of Michael Taylor*. $500-550.

Fig. 14-20. An absolute stunner! It is almost certainly Pennsylvania and beautifully made in every respect. Large squirrel cage swifts are relatively rare. *Courtesy of David Pennington*. $450-500.

Fig. 14-21. A wonderfully turned squirrel cage swift, probably Delaware in origin. The finest turnings and finish imaginable! It is somewhat smaller than the Pennsylvania squirrel cage swifts which it otherwise resembles in style. *Courtesy of Jeanne Asplundh*. $350-400.

Fig. 14-22. A nice restrained New England squirrel cage swift. It is noticeably smaller than the two Pennsylvania examples because it only needed to handle a 72+" skein and not the 90+" skeins from the larger "linen" reels. *Courtesy of Michael Taylor*. $275-300.

Fig. 14-23. A beautiful squirrel cage with two sets of barrels to make the handling of multiple skeins easier. Some with wide barrels had holes drilled in the horizontal supports so that dowels could be inserted to separate the skeins. *Courtesy of Jeanne Asplundh*. $400-450.

Basket swifts commonly found in New England testify to the utilitarian nature of textile tools (Fig 14-24). They are functional, but usually plain. Sometimes swifts are combined into a single piece with quill or bobbin winders.[8] Some of these winders are quite nicely crafted, but most are still pretty plain (Figs. 14-25 and 14-26). These were typically used by weavers in their work. Spinning wheels, winders and swifts were often in the home environment. They could show the family's wealth and taste. Quill or bobbin winders were not typically in that setting.

Fig. 14-25. A nice example of a New England bobbin winder with an accelerating head. Most of these are pretty plain but there are some Pennsylvania examples which are quite stunning. *Courtesy of Jeanne Asplundh*. $200-225.

Fig. 14-24. This New England basket swift is missing three of its four outside stays which would be easy to replace since one original remains. Dave's brother, Park, referred to all our treasured textile tools as "wooden get-in-the-ways". In the case of these utilitarian basket swifts, he was right on. *Courtesy of David Pennington*. $75-100.

Fig. 14-26. Another odd accelerated New England bobbin winder. *Courtesy of David Pennington*. $200-225.

One of the most valued and valuable of textile tools are the so-called tape looms. Their small size and relative scarcity make any a find, but many are also beautifully made. From their occasional appearance in formal portraits in the late 18th century, it is clear they occupied a special place in the lives of wealthy women.[9] Because few books show these looms, we have included more than a few examples.[10] It is a whole collectible area unto itself. Tape looms are sometimes referred to as rigid heddle looms because the heddles do not move. The plainest versions are simply flat boards with alternating holes and slots cut in them (Fig. 14-27). Usually these boards are shaped such that they can be easily held between the legs as the weaver uses them. Small tapes, an inch to an inch and a half wide, suitable for trim, edging, boarders or just plain ties for sacks, were woven on them. A few of these are flat boards big enough to sit right on the floor, with many having some sort of support or base (Fig. 14-28). Occasionally, however, one finds what appears to be a tape loom for a giant, but these were just meant to rest on the floor with a little support from the weaver's legs rather than in the lap between the legs. Wallace Nutting in his book, *Furniture Treasury*, shows a Pilgrim century chair with a solid back into which a later generation cut a tape loom![11] Of course, you couldn't sit in it and use it simultaneously, but by cutting into an old chair a frugal 18th century New England family created a tape loom of a convenient height.

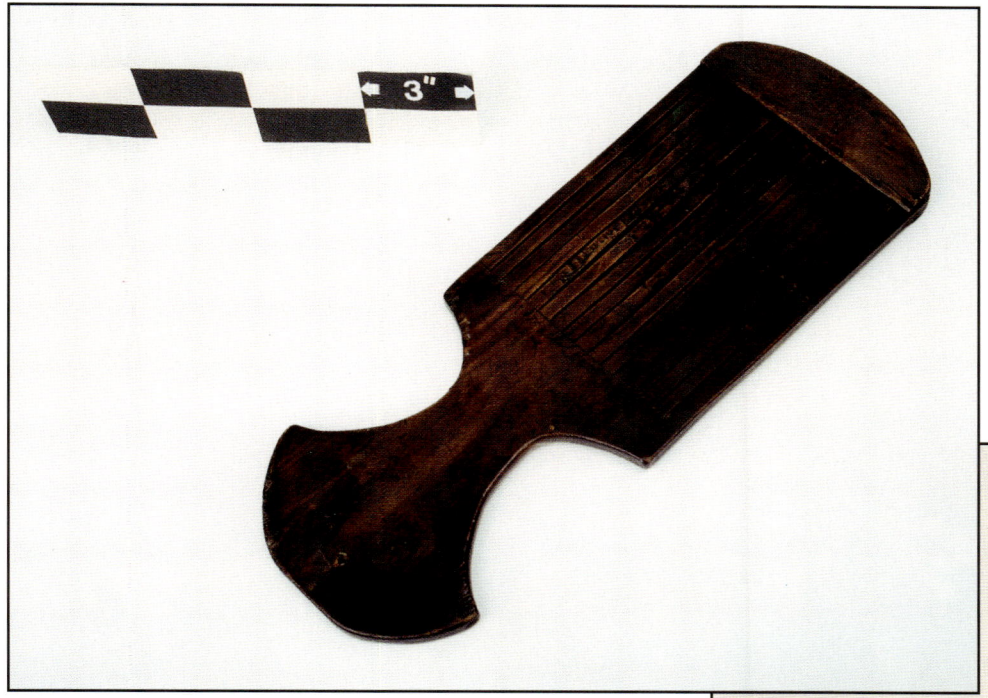

Fig. 14-27. A very nice example of a tape loom made extraordinary by an inscription on it in ink. "March 10, 1769 Mel Merrill made this Inkel Lome". Often referred to as Inkle looms, these are held between the knees and hence the cut out. *Courtesy of Jeanne Asplundh*. Without an inscription $300-350, with an 18th century inscription $900-1000.

Fig. 14-28. In spite of their plainness, these floor standing models of tape looms are quite desirable! *Courtesy of Jeanne Asplundh*. $500-600.

Fig. 14-29. Possibly from New England this box loom has great dovetailing and wonderful little lollypop decorations above the rigid heddle. These looms are delightful in their details and probably were not produced in quantity like the spinning wheels. *Courtesy of Jeanne Asplundh*. $900-950.

Fig. 14-30. These tape looms are interesting with their little hand treadles that make the harnesses go up and down like some kind of miniature. There are two or three examples with solid Shaker provenance, but this type seems to have been made by a variety of producers. Almost certainly late 19th century and still a bit of mystery as to origins. *Courtesy of Michael Taylor*. $550-600.

Box looms have become quite desirable in recent years with exceptional examples bringing two to three times the money that comparable spinning wheels would bring in spite of their relative simplicity (Fig. 14-29). Some are from fine furniture woods. It is these looms which often appear in formal portraits. Others are painted and decorated and are masterpieces of folk art.[12] Another style without a box, has movable heddles, little finger treadles and looks like a miniature table loom for a child (Fig. 14-30). These looms with their thick metal heddles give the appearance of a late 19th century catalog item but are earlier and widespread in New England. Some are solidly attributed to the Shakers, but most were made by others.

In England traveling box looms were developed. They contained everything inside a box which when unpacked became part of the loom. Some are quite elaborate, and at least one example appears to be American (Figs. 14-31, 14-32, and 14-33).

Other textile tools collected as accessories by wheel collectors include wool cards, wool combs, sheep shears, ripples, flax hackles, and flax brakes (see the photos in the "Gallery" at the end of this chapter for examples of all these tools).[13] Serious collectors usually succumb at least once to the temptation to own a full-sized loom, but one of those is more than enough unless you are going to use it.

---

[1] Laurel Ulrich, *The Age of Homespun*. (New York: Alfred A. Knopf, 2001): 187.

[2] Ibid., 289.

[3] The best discussion of the complexity of yarn count systems in Great Britain and America and their relationship to reel sizes is David Jeremy's "British and American Yarn Count Systems: An Historical Analysis," *Business History Review* (autumn 1971), 336-368. For additional information and measurements see Alan Raistrick's "Some Thoughts on Reels," *The Spinning Wheel Sleuth*, 15 (January 1997): 12-14. Although her discussion of how long it takes a reeler to wind off a bobbin load of yarn is fairly accurate, Ulrich's discussion of the yarn count systems is hard to follow.

[4] Jeremy, 347-348.

[5] Henry H. Taylor, "Some Connecticut Yarn Reels," *The Magazine Antiques* (June 1930): 538-540. Taylor was a very knowledgeable source and the only early expert who took reels seriously.

[6] Jerry Grant and Douglas Allen, *Shaker Furniture Makers*. (Hanover, N.H.: University Press of New England, 1989), 132-135. Elder Thomas Damon was in charge of table swift making from the 1840s to 1870s and averaged 920 a year from 1854-1860 at a price of 50 cents each. By 1854 he had constructed a machine for planning and edging swift slats. Damon refers to three sizes of swift in his 1854-1860 records, but Mike Taylor has found there were at least five. They come with cup or ball finials, and a few have a metal base instead of a wooden base.

[7] William Leinbach, "Daniel Danner: The Man Behind the Wheel," *The Spinning Wheel Sleuth* 4, (January 1994): 6.

[8] Florence Feldman-Wood, "Three Unusual Bobbin Winders," *The Spinning Wheel Sleuth* 36 (April 2002): 6-7 shows three unusual combinations of swifts and quill or bobbin winders from the American Textile History Museum.

[9] There is a formal portrait by John Singelton Copley painted around 1772 of Governor Thomas Mifflin and his wife with her using a tape loom to make fringe. It is owned by the Historical Society of Pennsylvania in Philadelphia. Elizabeth Haynes, "Mrs. Mifflin's Fringe Loom," *The Art of the Weaver* edited by Anita Schorsch, (New York: Universe Press, 1977): 28-29.

[10] Evelyn Neher's book, *Inkle*, has 50 page section on antique tape looms of all kinds. It also shows antique linen tapes produced from such looms and explains how to weave them.

[11] Wallace Nutting, *Furniture Treasury*, (New York: The Macmillian Co, 1928) Fig. 1788.

[12] The Philadelphia Museum of Art has a dated 1795 Bucks County, Pennsylvania box loom with painted decorations by John Drissel for Elizabeth Drissel which is an extraordinary piece of folk art. Irwin Richman, *Pennsylvania German Arts: more than hearts, parrots, and tulips*, (Atglen, Pa: Schiffer Publishing, Ltd., 2001): 80.

[13] Marion Channing's *Textile Tools of Colonial Homes* has wonderful flow charts of the uses of these tools as well as line drawings and explanations of the uses of each tool.

[14] Christopher Monkhouse. "The Spinning Wheel as Artifact, Symbol, and Source of Design." In *Victorian Furniture: essays from the Victorian Society autumn symposium*, edited by Kenneth L. Ames. Philadelphia: The Society, 1982:163-166 has a brief history of the spinning wheel chairs and some pictures.

Fig. 14-31. When closed up, this box loom with all sorts of tools and functions does not look like much, but stay with us as we open it up. Maybe they should be called "textile tool boxes". Most of these elaborate traveling weaving boxes are English. This one was found in a barn in New England, but only a wood test would be definitive. *Courtesy of Jeanne Asplundh.* $4000-5000.

Fig. 14-32. The "textile tool box" all set up for weaving. *Courtesy of Jeanne Asplundh.*

Fig. 14-33. The whole panoply of textile tools contained in the textile tool box. Truly an extraordinary piece of craftsmanship. They could be taken on short trips to a friends for a social visit or on a long trip to pass time. *Courtesy of Jeanne Asplundh.*

What follows is a "Gallery of Accessories". These accessories are not inferior to the ones used to illustrate the text.

Gallery of Spinning Accessories

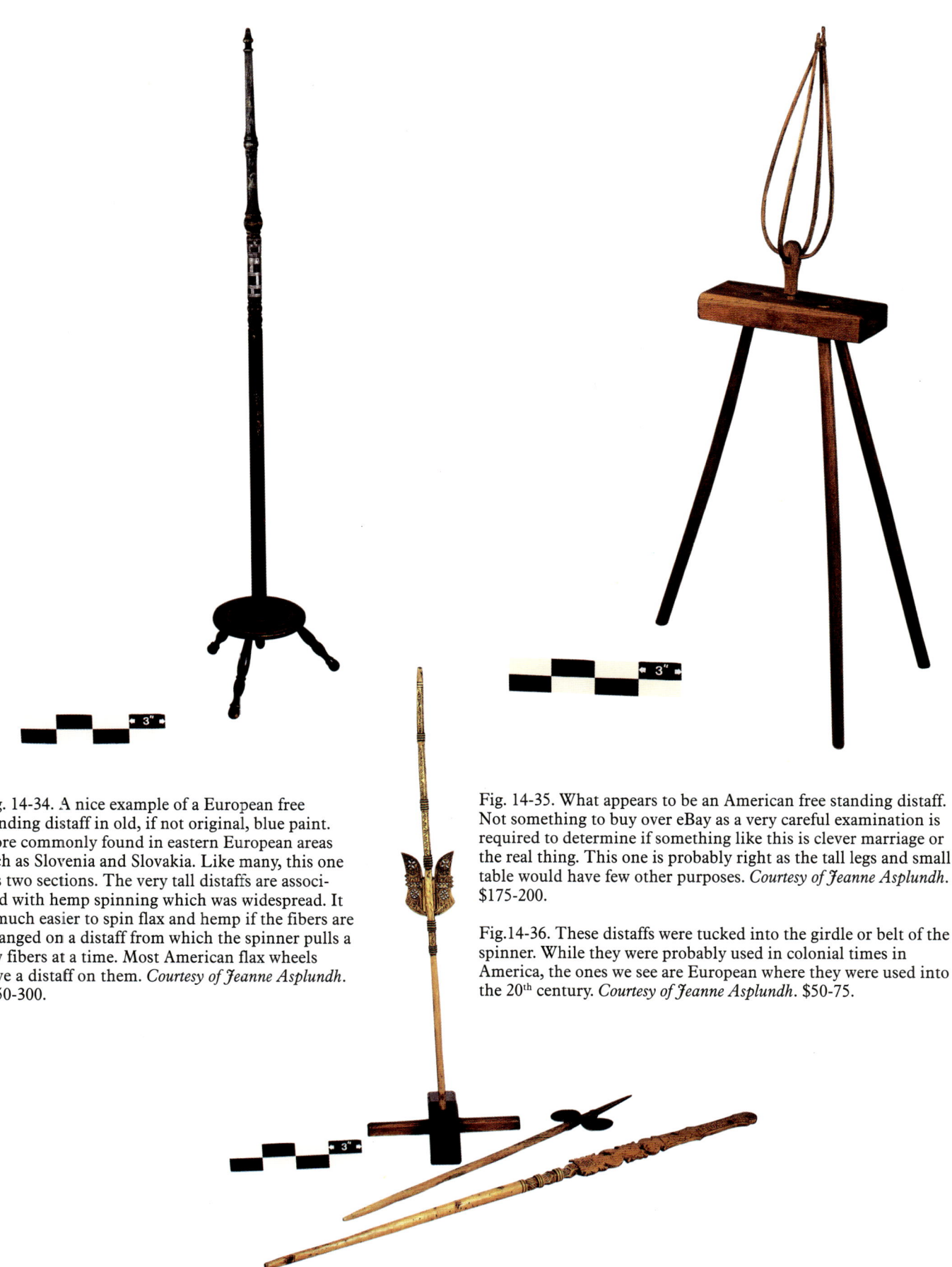

Fig. 14-34. A nice example of a European free standing distaff in old, if not original, blue paint. More commonly found in eastern European areas such as Slovenia and Slovakia. Like many, this one has two sections. The very tall distaffs are associated with hemp spinning which was widespread. It is much easier to spin flax and hemp if the fibers are arranged on a distaff from which the spinner pulls a few fibers at a time. Most American flax wheels have a distaff on them. *Courtesy of Jeanne Asplundh.* $250-300.

Fig. 14-35. What appears to be an American free standing distaff. Not something to buy over eBay as a very careful examination is required to determine if something like this is clever marriage or the real thing. This one is probably right as the tall legs and small table would have few other purposes. *Courtesy of Jeanne Asplundh.* $175-200.

Fig.14-36. These distaffs were tucked into the girdle or belt of the spinner. While they were probably used in colonial times in America, the ones we see are European where they were used into the 20th century. *Courtesy of Jeanne Asplundh.* $50-75.

173

Fig. 14-37. These are "drop spindles", the original spinning device. The twist is created by the rapid rotation of the spindle by hand. Shepherds tending flocks of sheep could spin while following their sheep. Of course, in poor countries a drop spindle was an affordable way to produce yarn. Peasants did so well into the 20th century. *Courtesy of Jeanne Asplundh.* $5-10.

Fig. 14-39. This wicked looking device is a flax hackle. It was used in preparing flax for spinning. Sometimes, there are more than one set of teeth mounted on the same board with finer teeth set nearer together. Often there is a wooden cover for protection when not in use. Usually rusty and nasty and not of much value ($25-30), but signed and dated ones in good condition are nice. This one in tiger-striped maple and stamped "F.A. Goodrich" is special! *Courtesy of David Pennington.* $100-125.

Fig. 14-38. A cute little holder for something! Here we are frankly speculating, but it is possible that this stand was used with spindles (reproductions shown in picture) as a plying device. *Courtesy of Jeanne Asplundh.* $35-40.

Fig. 14-40. Wooden spikes instead of nails are occasionally seen. Probably used as hackles but possible ripples for removing flax seed. *Courtesy of David Pennington.* $90-100 each.

Fig. 14-41. A ripple for removing the flax seed prior to preparation for flax preparation for spinning linen. Flax seed was gathered and pressed into linseed oil. It was a very important commercial product and meant that the laborious tasks of growing and processing flax yield two important raw materials and not just flax for linen. Some ripples are cast iron and have tangs at each end, such that they can be driven into a stump or block of wood. *Courtesy of David Pennington.* $75-90.

Fig. 14-42. More nasty looking metal implements! A "pair" of metal combs. Metal combs did not come in pairs although they were used two at a time. Long-fibered wools were combed by highly skilled craftsmen who heated a comb which was then mounted on a special post (hence the extra hole in the handle). A second comb was then used to pull the fibers straight for spinning worsted yarns. This craftsman was the best paid of all manual textile workers. He would typically have several combs heating on a small specially designed stove. (See Eliza Leadbeater's book, *Spinning and Spinning Wheels*, for pictures of someone doing the process). *Courtesy of Jeanne Asplundh.* $75 each.

Left:
Fig. 14-43. Wool or cotton cards. These do come in pairs and were used to card wool (sometimes cotton) by hand. Used by handspinners throughout the 19th century, it was a laborious job well-suited to keep children occupied and productive. Early in the 19th century large, water-powered, carding machines were developed. Fleeces could be taken there and processed, the carder often keeping a portion of the processed wool in payment. New ones are still available. *Courtesy of Jeanne Asplundh.* $30-35 per pair if old and nice.

Fig. 14-44. A spool knave or lazy kate. A bobbin could be loaded on each wire and then they would turn freely during the plying process. Always a good find for a textile tool collector. *Courtesy of Jeanne Asplundh*. $100-125.

Fig. 14-46. A very nice New England squirrel cage swift with some nice decorative details. This one shows some more Yankee ingenuity. The top cage is fixed. The bottom cage is free to travel up and down. That allows gravity to provide the appropriate size and obviates the need for any threaded screws! *Courtesy of Jeanne Asplundh*. $275-300.

Fig. 14-45. Another very nice tall Pennsylvania squirrel cage swift. The Pennsylvania examples are taller than the New England examples because of the larger skeins they handled. The former are typically also much more stylish. *Courtesy of James Munsie*. $275-300.

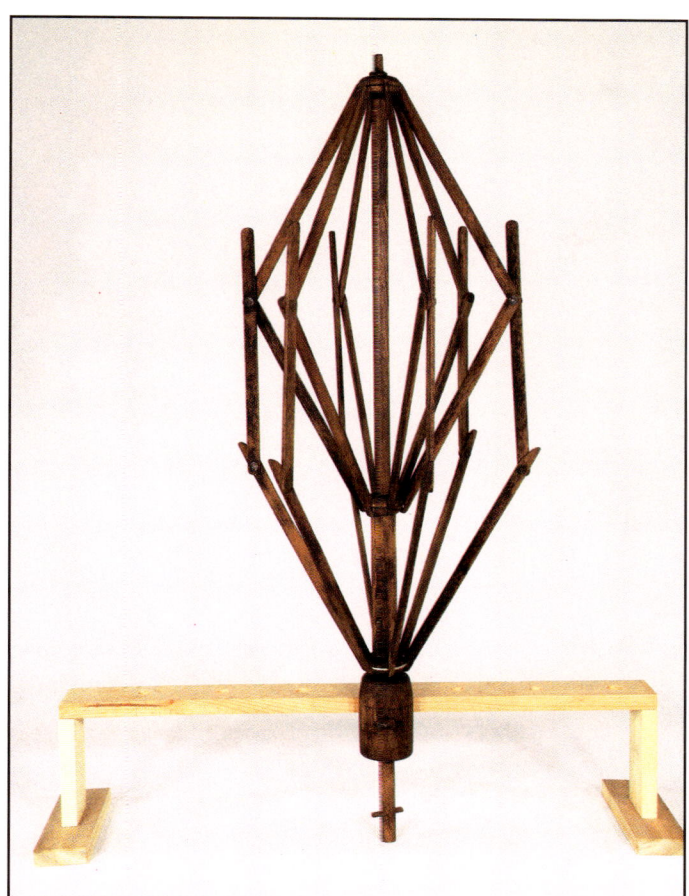

Fig. 14-47. Mike purchased this unusual umbrella swift in northern Michigan for Dave. It was labeled a "noodle dryer"; maybe it was, but we think not. It is still known as a "noodle dryer" type although we have seen no others. Rarity does not always make for value. *Courtesy of David Pennington.* $75-90.

Fig. 14-48. It looks very old and is probably English. If American, wow! It is certainly a stunning example of an umbrella swift. *Courtesy of David Pennington.* $200-225.

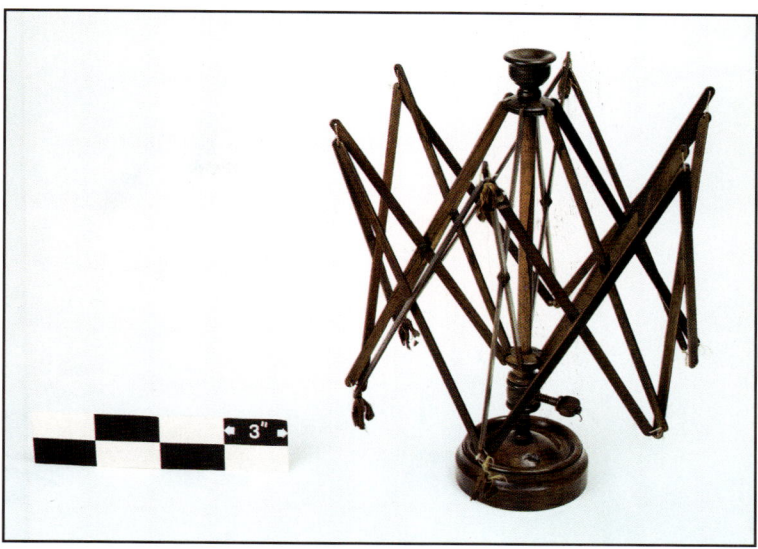

Fig. 14-49. A lovely, diminutive table model umbrella swift. *Courtesy of Jeanne Asplundh.* $150-175.

Fig. 14-50. A cute little squirrel cage swift with nice bone rods for the cages. Probably a European fancy for women's embroidery work. $1100-1250.

Fig. 14-51. Small separate squirrel cages from this size on down to the teeny, tiny thumbnail size (in ivory or bone) were used in embroidery work. This pair is English and in metal. Sometimes only one of a pair is found which diminishes the value dramatically. *Courtesy of Jeanne Asplundh.* $250-275.

Fig. 14-52. No this reel is not broken. Many Pennsylvania reels have one hinged arm that allows for skeins to be easily removed. There is a sleeve that holds the arm straight when it is reeling. (Caution: some of these arms are not attached and will drop with a thud, so when removing a sleeve for inspection, support the end of the arm.) A much simpler way to remove the skein is to have one cross arm without a lip on it (Fig. 14-53). Notice also that this reel has a single rear leg which makes it fit better into corners than against a flat wall. Most three legged winders have two rear legs and one front leg. This reel has the double dial feature which accounts for its value. *Courtesy of Michael Taylor.* $350-400.

Fig. 14-53. This side view of a typical six-armed reels shows that on most reels one of the cross arms has no front flange, rather it is rounded off. That allows the reeler to remove the skein by simply pulling the skein off of that arm. *Courtesy of David Pennington.*

Fig. 14-54. At first glance, just another four-armed click reel, but this reel has a mechanism to keep the gears from being damaged if turned clockwise repeatedly. *Courtesy of David Pennington.* $200-225.

Fig. 14-56. Here we see what happens if the winder is turned incorrectly. The whole slapper mechanism pivots out of the way and no harm is done. Bill Ralph saw one with this type of mechanism signed "J. Farnham". *Courtesy of David Pennington.*

Fig. 14-55. Here we see what happens if the winder (Fig. 14-54) is turned correctly, counterclockwise. The slapper will make a noise as it returned to rest after being pushed away by the rotating peg attached to the toothed gear. *Courtesy of David Pennington.*

Fig. 14-57. Another example of the rare and desirable, slant tabled six-armed Connecticut reel (See Fig. 14-6). This one has a superb wooden arrow on its dial and is in superb condition except for minor worm damage in one leg. A series of freezing and thawing rounds in a deep freezer – two weeks in, two weeks out and two more weeks in - handles most beetle infestations. If not, see an expert quickly. *Courtesy of Michael Taylor.* $250-300.

Fig. 14-58. A very nice example of a typical New England style of reel. With exposed gears the user can simply watch the peg go around, and no separate dial is needed as in the case of the closed box reels. These typically have forty teeth in the gear and the worm gear is cut so that it takes one revolution of the arms to move the gear ahead one tooth. 40 teeth, forty revolutions, two yards in circumference, and one click is 80 yards. *Courtesy of David Pennington.* $150-200.

Fig. 14-60. Anything in old paint is good and most wheels and reels are not in old paint. The Berks County, Pennsylvania area is becoming known for its striped wheels and reels. Here is a nice example of the kind of banding encountered. Notice also that there is one front leg and two rear legs making this reel more suitable for a wall space than a corner space. *Courtesy of Michael Taylor.* $250-300.

Fig. 14-59. Another reel of the open geared New England variety, but this one is as exciting as the previous example is pedestrian. Great base similar to Fig. 14-1. The handle is a nice extra touch. *Courtesy of David Pennington.* $225-250.

Fig. 14-61. An elegant reel of the sort highly desired by collectors. Probably Connecticut. *Courtesy of David Pennington.* $250-300.

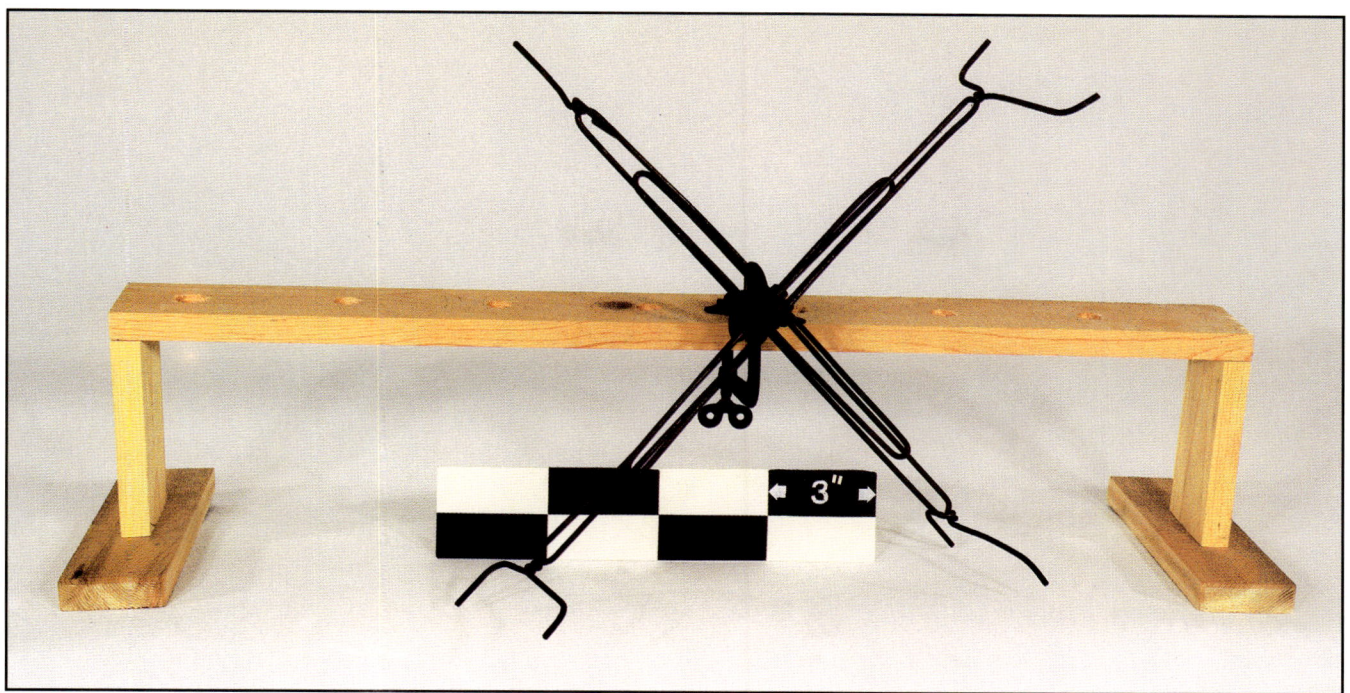

Fig. 14/62. From the elegant to the industrial. This little metal reel looks like a bunch of coat hangers, but once unfolded it has a little clicking mechanism which sounds off after 40 revolutions just like the big wooden ones. Patented in 1867 it occasionally shows up in flea markets. *Courtesy of David Pennington.* $50-60.

Fig. 14-63. Another example of the table tape looms with two harnesses. If these were not so widely represented in well-known museums and if some were not solidly attributed to the Shakers, this group of looms might well be thought of as a late 19th century catalog item, but they seem to be an earlier form. Having only two harnesses and four to six heddles, they could be used only for very simple, narrow strips, perhaps cords for sacks. *Courtesy of David Pennington.* $550-650.

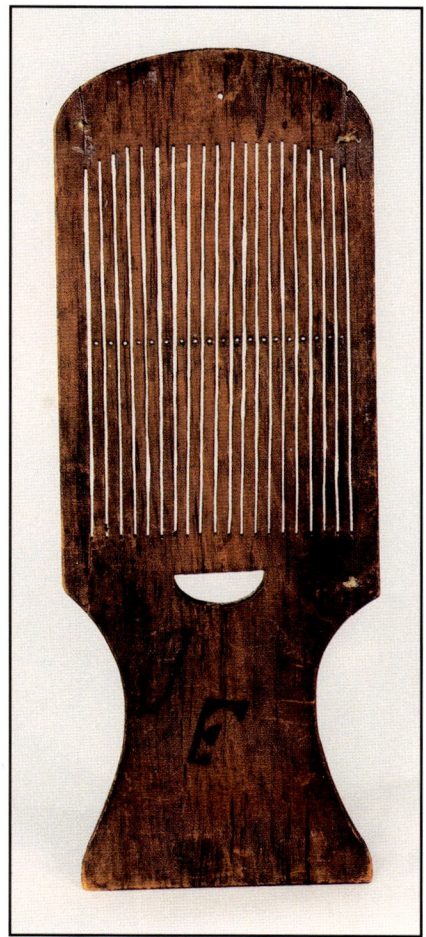

Fig. 14-64. A wonderful example of a basic tape loom. Some are initialed or dated which adds to the value. *Courtesy of David Pennington.* $275-325.

Fig. 14-65. A very nice floor standing tape loom. Some of these look like the extremely large tape looms which were not held between the knees but between the calves with the butt end resting on the floor. Some of them may have been converted with the addition of a stabilizing piece in the front which would allow it to stand freely when not in use. *Courtesy of Jeanne Asplundh.* $600-700.

Above right:
Fig. 14-66. A primitive, but wonderful, box loom on legs with treadles. In some cases a relatively nice box loom sits on a primitive set of legs and treadles, which suggests an early improvement. In this case it is hard to tell if the original maker put on the legs or if they were added early on. *Courtesy of Jeanne Asplundh.* $1500-1600.

Right:
Fig. 14-67. In this case there is no question that this small loom was made as we see it. This might be referred to as a child's loom, but it is quite suitable for someone doing a lot of serious inkle weaving. *Courtesy of Jeanne Asplundh.* $700-800.

Fig. 14-68. Another example of the fancy box loom found in Europe. It is made of mahogany. In this case the rigid heddle cannot be removed and stored, which makes it very vulnerable to breakage if carried about. This one probably stayed at home, but if suitably protected it could travel. *Courtesy of Jeanne Asplundh*. $800-850.

Fig. 14-69. The previous box loom with the lid over the little spools open. The lid has holes in it, so the threads from the spools can be pulled through while creating a warp. *Courtesy of Jeanne Asplundh*.

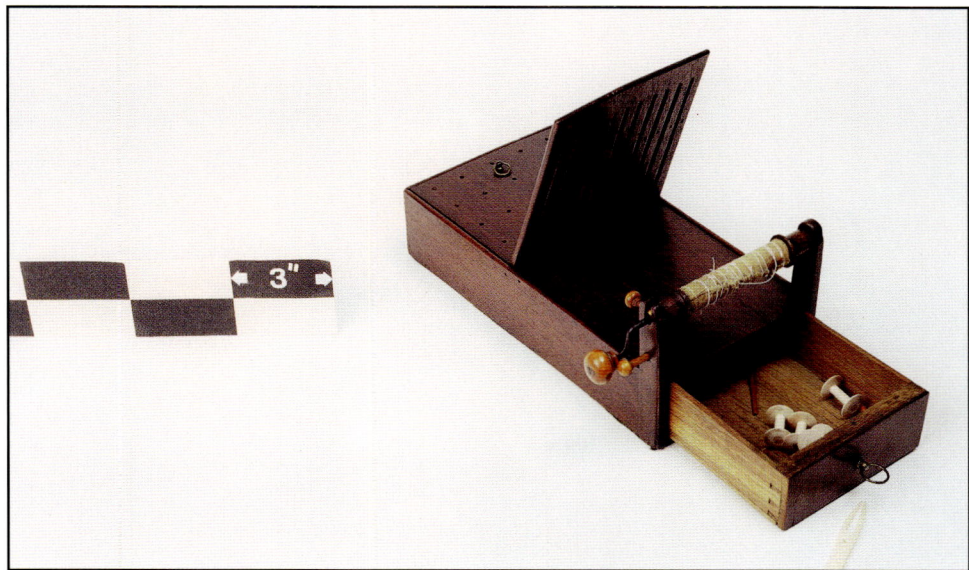

Fig. 14-70. The same loom showing the drawer open. Small weaving accessories, such as combs, could be kept there. This example has an understated elegance to it. Unlike the "textile tool boxes" with dozens of little accessories, this one is less gift-like and more practical. *Courtesy of Jeanne Asplundh*.

Fig. 14-71. Wonderful red surface on a Pennsylvania box loom. There is no cloth beam for the finished tape. The cross-arms on the warping beam are one way of expanding the beam's circumference. On this example the top piece of the rigid heddles is pegged on. Some box looms have no top cross piece. *Courtesy of Michael Taylor.* $750-800.

Fig. 14-72. Another type of Pennsylvania box loom with an internal cloth beam as well as a warp beam. Extraordinary attention to detail throughout with dramatic use of quartersawn oak. *Courtesy of Michael Taylor.* $850-900.

Fig. 14-73. From the estate of Eleanor Freeborn, a famous collector of Pennsylvania textile tools, this tape loom has the heart motif cutout in one end of the box. Quite possibly a special gift item, it is another superb example of this form. *Courtesy of Jeanne Asplundh.* $900-1000.

Fig. 14-74. This box loom has its two moveable wooden heddles between the cloth beam and the warp beam. The cloth beam is on the front left and would hold the woven tape. The warp beam in the back right would hold only warp threads and would unwind during the course of the weaving as the finished tape was wound onto the cloth beam. In a sense this type of box loom is simply a miniature two harness loom. *Courtesy of Jeanne Asplundh.* $1100-1200.

Fig. 14-75. A close up of the box loom in Fig. 14-75 showing the two wooden heddles which could be alternately pulled up by hand to create the shed for weaving. *Courtesy of Jeanne Asplundh.*

Fig. 14-76. Probably English, possibly 18th century, this oak box loom has a detachable spool holder for warping. *Courtesy of Jeanne Asplundh*. $1500-1600.

Fig. 14-77. A second look at the English box loom in Fig. 14-77 showing the separate pieces. Once warped the spool box could be removed to make weaving easier. *Courtesy of Jeanne Asplundh*.

Fig. 14-78. A fancy table top reel from Europe. It is double geared and has a drop arm arrangement to allow for the easy removal of a skein. *Courtesy of Jeanne Asplundh*. $500-550.

Fig. 14-79. A horizontal squirrel cage swift with a spool winder and storage area around the base for baleen spools. It is a beautifully crafted item out of fine materials, probably used with silk skeins. *Courtesy of Jeanne Asplundh.* $900-1000.

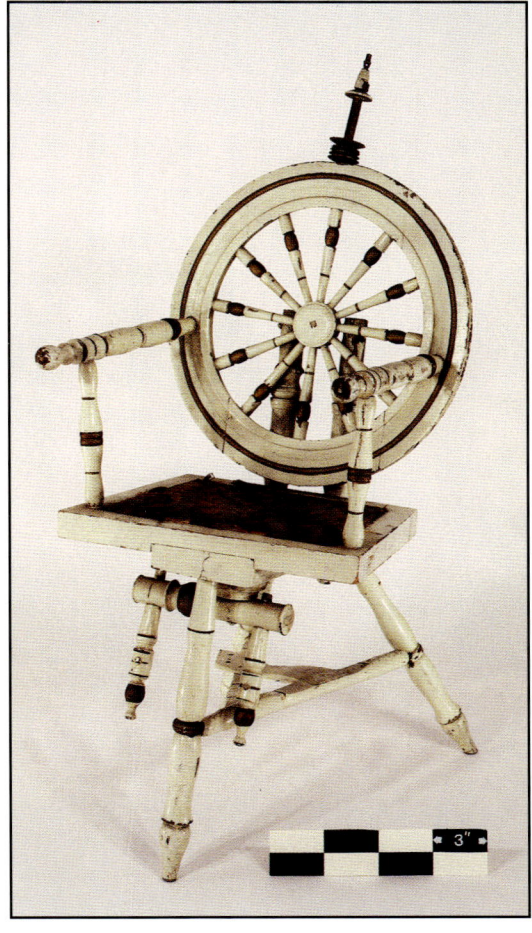

Fig. 14-81. Not really an accessory, but unfortunately the fate of some spinning wheels in the late 19th century when spinning wheels had become a curiosity.[14] *Courtesy of David Pennington.* $175-225.

Left:
Fig. 14-80. This little American table swift was once thought to be Shaker because of its clean lines and ingenuity, but lots of New Englanders valued both. Here it is set up for use as a swift. The arms can extend out to hold the skein of yarn, and the arms will rotate to allow for the unwinding process. By unthreading the central cup, the arms can fold back side by side and retract for easy storage. Definitely a clever little item which may have been made and sold to the resort trade by the Shakers in the late 19th century, but we have no evidence. *Courtesy of Michael Taylor.* $150-200.

Chapter 15

Tips on Collecting

The spinning wheels shown in this book come from four notable private collections. The wheels at the American Textile History Museum came primarily from three great collectors: S.D. Stevens, Anne Pixley, and Joan Cummer. All of these seven collections had (has) thirty or more wheels, but each of the seven collectors had a somewhat different philosophy as well as opportunity for collecting. Collectors are not thinking first about resale value. They are thinking about improving their collection.

The Connecticut chair wheel marked "J. Miles" (See the index for photographs of the makers mentioned in this chapter.) is not in good condition. It has been extensively restored and the finish on it is poor. It is, however, extremely rare to find a signed Connecticut chair wheel, and this example is in the best condition of those known. So to a collector it is a find. The double-flyer wheel marked "J. Miles" is in wonderfully complete and "as found" condition with a superb finish. It is one of the most desirable types of double-flyer. In the antique market today, which values completeness and original finish, the double-flyer wheel is the more valuable wheel.

Most people who get involved in collecting America wheels are interested initially in a representative collection, so they look for a nice saxony, a good wool wheel, a chair wheel, a double-flyer wheel, and an Irish castle wheel. Condition, completeness, and finish are very important. Warped drive wheels, which will not hold a drive cord, are bad news. Missing bobbin and flyer units or replaced ones which fit poorly are a big liability. The original finish is a big plus.

If the wheel is marked by a famous maker, such as Joel Farnham, Daniel Danner, Alpheus Webster or Beardsley Sanford, that is a definite plus. If it has known initials from a Shaker trustee, such as "SR AL," that is also good. See what you can learn about the maker of a marked wheel. Use genealogical records, census records, local tax records and the local historical society. It was fascinating to read Samuel Humes' probated inventory at the Lancaster Historical Society in Lancaster, Pennsylvania and learn that he still had unfinished wheels and reels on hand when he died at age 83.

Once the collecting fever takes hold, it can go as interest, luck, or circumstance dictate. Eastern Pennsylvania is a great collecting location as is upstate New York and most of New England, but your collection will look quite different if you are buying mostly local wheels.

Study your wheels carefully and look for little details. The Shaker wheel makers at Alfred, Maine, put little sets of dots both by the orifice of the flyer and metal nut in the flyer whorl to keep the metal pieces matched up while the wooden pieces were being affixed. "S. Reiner," a Pennsylvania maker, marked the three wooden pieces of his bobbin and flyer units with Roman numerals. We assume that was because after they were assembled, he disassembled them to put a stain on the pieces, and he didn't want to get the different sets of pieces confused. The big patent wheels were disassembled for shipping, so the pieces were sometimes given a number. Examples of such numbering can be found on T.D. Brown wheels, Lyman Wight wheels and G. Hathorn wheels. The bottom of treadles and undersides of tables sometimes have marks or dates in pencil. The insides of click reels may be dated in pencil.

One thing we noticed in the study of saxonies from

eastern Pennsylvania is that having the wheels in front of you makes all the difference in noticing small, but significant, features. Examining pictures is very useful for picking up differences in turnings, numbers of spokes, use of color, and tensioning arrangements. With the wheels directly in front of us, we started noticing how the drive wheel was made, a bead on the lip of the drive wheel, how the wheel support posts were fastened, how the treadle diagonal piece was attached to the treadle bar, how the diagonal piece was decorated, and which way it tapered. All of these features were consistent within a small group of makers, and it made us more observant of other wheels later.

Study the patterns of wear on wheels. Some wheels were treadled with both feet as a study of treadle wear shows. Look at the wear marks caused by the thread as it passed over the wooden and metal parts of a bobbin and flyer unit. It is clear that one way of slowing down the intake on a flax wheel was to wrap the thread around to the backside of the orifice which left telltale marks.[1]

Most wheel collectors also collect related textile tools. Reels, or winders are measuring devices which look a little like windmills. Since they were often made by spinning wheel makers, a signed winder that matches a signed wheel is a great find. Niddy noodies are primitive winders and are a must. Some are elaborately chip carved. Swifts are "unwinders" and come in several designs, but a spinning wheel collector will want a good example of an umbrella and a squirrel cage swift. A well-to-do collector will want a scrimshaw swift, but they cost more than almost any wheel.

In addition to this book a collector must have a sound knowledge of how spinning wheels work. Otherwise, you may miss obvious marriages of wheel parts, which make the wheel, not a collectible mystery wheel, but a piece of junk. Truly great, complete wheels are seldom found. So in addition to knowing how to spin, a collector will soon acquire a wheel repairman. They aren't in the phone book, and the person needs to be a turner of wood and metal, a blacksmith, adept at working in horn and ivory, and a refinisher. Not surprisingly, most people settle for a good wood turner, avoid problems that require metalwork and do their own staining. Go look at major collections, the more wheels you see, the more you will know what you are seeing. Subscribe to *The Spinning Wheel Sleuth*. Get copies of Joan Cummer's and Patricia Baines' books.

All of us who collect have made bad buys and also have "passed" on items we wished we had purchased. That is part of collecting. It helps if you work with a dealer who will take something back, but that option is not always available. Buying sight unseen from long distances is obviously risky, but we know serious collectors who do with both good and bad results. One important variable we haven't mentioned is storage space. Having an extra barn or two is a big plus for the serious collector. Of course, there is no right way to collect. Collecting spinning wheels or anything else is a passion. Enjoy yours.

---

[1] Michael B. Taylor, "Deciphering Wear Marks," *The Spinning Wheel Sleuth* 18 (October 1997): 5-7.

Appendix

Spinning Wheel Makers: Marks and Information

The following list is of makers whose wheels we have either seen or encountered in our research over the last nearly forty years. Dave has kept a file of every marked wheel during that time and finally transferred some of the information to a database in order to make searches easier. The pictures and rubbings of the marks couldn't be included but some basic information may prove useful to the reader. As it was initially just for our use some of the descriptions may need a little explanation to be helpful to anyone trying to track down a maker or mark found on a wheel. The maker's name or initials are shown in capital letters as almost no marked American wheels have been seen so far with lower case letters. The name is then shown in an alphabetical list assuming the second initial represents the last name. The periods between initials are sometimes there but have been assumed when not present.

The "**Type**" column tells what variety of items we have encountered followed by a "**Seen?**" column that lets you know if this is a surviving artifact or information encountered through research. "**Dates**" may be either working years or the lifetime of the maker. The "**Location**" entry showing the state and city, if known, is followed by the "**Certain?**" entry. "**Certain**", "**possible**" and "**probable**" are the three choices based upon what we knew about the wheel's provenance, style or general characteristics. "**Comments**" contain additional information that may be useful and uses some of the following abbreviations:

"*": An asterisk in the comments column means this maker's wheel is pictured in the book.

"**AWF**": *American Windsor Furniture* list of chair and wheel makers.

"**A.T.H.M.**": The American Textile History Museum

"**SWS**": From the *Spinning Wheel Sleuth* listing of wheel makers. Additional data from this list follows the "SWS" initials.

"**Winder**": Reel and winder are terms we have used interchangeably.

"**Esposito**": Ralph Esposito had a list of wool wheel makers in his thesis.

"**Flax**": Implies a Saxony-style wheel unless there is additional information in the comments

"**Type 1 double flyer**": The most often found style of double flyer wheel. See Fig.7-1

"**Type 2 double flyer**": The "t- top" style of double flyer shown in Fig. 7-16

"**Type 3 double flyer**": A vertical double flyer wheel with the wheel above the flyers. Fig.7-12

"**wh.**" wheel

Cryptic comments may be found which describe, at least to us, the style of the turnings on a leg, post or wheel spoke, i.e. "olive into a shotgun shell" (sgs). In this case the rule to follow is, if it doesn't make sense to you, forget it! It is, however, our hope that this listing will actually answer more questions than it raises, but you never know. As other similarly afflicted "wheel people" research makers and wheels, this list will become outdated and we're sure someone will look at it on the first day and see an error or omission. Dave takes full credit for all errors! The file was started many years ago and not every research breakthrough has been passed along to us. There are over a thousand makers and marks included here. Enjoy.

SPINNING WHEEL MAKERS

MARK	NAME	SEEN	DATES	TYPE	LOCATION	CERT	COMMENTS
A.A.	A, A	Yes		Flax	MD	Poss.	
E.A.	A, E.	Yes		Bat's head	MA, Hancock	Prob.	Shaker
1834 GLA N111	A, G.L.	Yes		Flax			SWS
1814 HLA N459	A, H.L.	Yes		Flax			SWS
I.A.	A, I.	Yes		Wool	NY	Prob.	
J.A.	A, J.	Yes		Flax	N.E.?	Prob.	E-Bay auction
L.A.	A, L.	Yes		Flax			plain wool
P.A.	A, P.	Yes		Flax, Wool	CT	Poss.	upright double flyer, fat flame finials, wheel cherry
T.A. 177	A, T.	Yes		Unusual			Mount Vernon, also Bill L., pie cut wheel, 16 spoke, beaded table
W.A.	A, W.	Yes		Flax	PA	Poss.	
DANIEL ABBOTT	ABBOTT, DANIEL 1749	Yes	1749	Winder			SWS
ADAMS	ADAMS	Yes		Flax			
G.P. ADAMS & CO.	ADAMS, G.P.	Yes		Flax			has # 18118 on it
JOHN ADAMS	ADAMS, JOHN	Yes		Wool	N.E.?	Prob.	has Miner's head: Oswego,NY with "Burlington"
S.A. ADAMS	ADAMS, S.A.	Yes		Wool			SWS
SAMUEL ADAMS	ADAMS, SAMUEL	Yes		Flax ?	NY	Poss.	Esposito, SWS, AWF
	AIKEN, THOMAS	No	1747-1831		NH, Deering/Antrim	Cert.	
AKIN + LYMAN	AKIN, + LYMAN	Yes		Flax			
L. ALBERT	ALBERT, L.	Yes		Winder	NY	Prob.	6 armed winder
P. ALBERT	ALBERT, P.	Yes		Winder	PA		6 armed winder
	ALBRIGHT	No					
H. ALDERFER	ALDERFER, H.	Yes		Flax			ALDEREER?
S. ALDRICH	ALDRICH, S.	Yes		Wool	NY	Prob.	
ALEXANDER	ALEXANDER	Yes		Flax	CT	Prob.	* "DER" not distinct
F. ALLEN	ALLEN, F.	Yes		Flax			E-Bay auction
	ALLEN, JAMES	No			NH, Walpole	Prob.	SWS
T. ALLEN	ALLEN, T.	Yes		Wool	MA	Poss.	"T" could be "F
	ALLEN, WARREN	No	1829,Aug15	Patent	NY, New Haven	Cert.	Patent missing - "Spinning machine"
	ALLISON, BURGESS	No	1813,Mar3	Patent	NJ, Burlington	Cert.	Patent missing - "Spinning machine"
	ALLISON, WILLIAM	No	1790-1794		PA, Phila.	Prob.	poss. wheel maker
	ALRICKS, JAC.	No	1809,Oct11	Patent	DE, Wilmington	Cert.	Patent missing - "Spinning machine"
	ALTER, JOHN	No	1812 prior		OH, Zanesville	Cert.	AWF
	ALTOFFER, J.	No	1821,Aug30	Patent	VA, Harrisville	Cert.	Patent lost,"Spinning wool & cotton"(See Brushnell)
DAVID ANDERS	ANDERS, DAVID	Yes		Flax	VT	Poss.	
	ANDERSON, H.	Yes			VA	Poss.	
J.A.	ANDERSON, JOHN	Yes	1806-1814	Wool	ME, SDL	Cert.	Shaker Trustee
	ANDRUS, E. PALMER	No	1849-1870	Flax, Wool	WI, Winooski	Prob.	SWS
	ARCHER, J.	No		Wool	NC	Poss.	
ARCONA	ARCONA	Yes		Wool			
	ARMSTRONG, JOSEPH	No	1799		PA, Greensberg	Cert.	AWF
ASHER-WILCOX	ASHER-WILCOX	Yes		Unusual			double flyer, prob. type 1,

SPINNING WHEEL MAKERS

N. AVERILL	AVERILL, NATHANIEL		No	1664-1751	MA, Topsfield	Cert.	making wheels IN 1721 wheel seen with "PARADICE"
B.B.	B, B.	Flax	Yes		CT	Poss.	
B.B.	B, B.	Unusual	Yes		NH	Poss.	1st "B" reversed, double wheels & treadles, 4legs, horiz. 2piece fram
B.B.	B, B.	Unusual	Yes		NH	Poss.	double flyer, type 1
B.B.	B, B.	Unusual	Yes				SWS- "double flyer and double wheel chair wheel"
B.B.	B, B.	Unusual	Yes		CT	Poss.	double flyer, type 1
B.M.B.	B, B.M.	Unusual, Wool	Yes		MA	Poss.	double flyer, type 1
CXB	B, C	Wool	Yes		VA	Poss.	side screw tension, half moon painted piece, legs out end of table
D.B.	B, D.	Flax	Yes		NH	Poss.	
d+B 4-78	B, D.	Flax	Yes		PA, Somerset Co.	Prob.	
E.B.	B, E.	Unusual	Yes		CT/PA	Prob.	* double flyer, type 1 ("Logs" tension), also type 2 (T-top)
H.M.B. 1814 501	B, H.M.	Flax	Yes				dated 1814, #501
H.W.B. 1649	B, H.W.	Flax	Yes	1649			
I.B. 1834	B, I.	Unusual	Yes				Swiss style, dated 1834
J.B.	B, J.	Wool	Yes		NY	Prob.	
J.B. 1856	B, J.	Flax	Yes		PA/OH	Prob.	See "BIRCHFIELD"
J.B.	B, J.	Unusual	Yes		NY	Prob.	Farnham style
L.B.	B, L.	Wool	Yes				stamp in fancy script
M.M.B. 1841 50?	B, M.M.		Yes	1841			SWS
S.B.	B, S.	Flax	Yes		CT	Poss.	
	BABB, THOMAS N.	Patent	No	1880	TN, Monroe	Cert.	see KEISLING
J. BABCOCK	BABCOCK, J.	Flax, Wool	Yes		VT, Bennington	Prob.	sax. based winder
S. BABCOCK	BABCOCK, S.	Flax	Yes				signed on side of sax.
	BADERTSCHER, PETER		Yes	1805-1872			Pat. type table model
B.B.	BAILEY, BENJAMIN	Wool	Yes	1799-1880	ME, Alfred	Cert.	Shaker Trustee at Alfred 1839-1869
G. BAIR	BAIR, G.	Winder	Yes				
H.S. BAIR	BAIR, H.S.	Winder	Yes				
D. BAKER	BAKER, D.	Flax	Yes		NJ	Poss.	SWS
D. BAKER	BAKER, D.	?	Yes		PA	Poss.	
	BAKER, SAMUEL		No	1780	NC	Poss.	SWS
	BAKER, SAMUEL		No	1870's	MA, Lowell	Poss.	poss. wheel maker
	BAKER, SAMUEL		No		NC	Poss.	poss. wheel maker, Esposito
WILLIAM BAKER	BAKER, WILLIAM	Winder	Yes				
	BALDWIN, JOEL	Unusual?	No	1790's	CT, Bristol	Prob.	ad- made S.W., reels, & double flyer
	BALDWIN, SYLVANUS	Patent	No	1812, May 6	VT	Cert.	Patent missing - "Spinning machine", see Town, also 1815, Aug18
M. BALL WARRENTED	BALL, M.	Flax	Yes		OH	Poss.	"WARRENTED" stamped on side of table
A. BALLOU	BALLOU, A.	Wool	Yes				SWS
A.B.	BANGS, ALLEN	Wool	Yes	1813-1822	ME, Poland Hill	Cert.	Shaker Trustee
S. BANTHAM 1826 No 4	BANTHAM, S.	Flax	Yes	1826			dated 1826 & "No.4"
J. BARGE	BARGE, J.	Flax	Yes		PA, Lancaster	Cert.	stamp "1000"
E. BARKER	BARKER, E.	Wool	Yes		ME, Alfred	Cert.	Shaker -stamped on signed SR AL wh.
	BARNES, TURNER	Patent	No	1867	IN, Greensburg	Cert.	Pat. #69387 IN 1867

SPINNING WHEEL MAKERS

	Maker				Location		Cert.	Notes
S. BARNUM	BARNET, SAMPSON			1779-1807	PA, DE	No	Cert.	Chester Co, PA1779-1781+1814-1823, Wilmington, DE-1789-1807
	BARNUM, S.	Wool, Unusual	Yes		CT/ NY		Poss.	* double flyer type 1, N.Y. Barrel tension
D. BARRET	BARRET, D.	Flax	Yes		CT		Poss.	dated 1806, diff. D. BARRET stamp
D. BARRET	BARRET, D.	Flax	Yes		CT		Poss.	dated 1786
	BATCHELDER, C.	Patent	No	1824,Jul3	NY, Lowville		Cert.	Patent missing - "Spinning machine" (SEE MERRILL & KING)
	BAUERLE, LEONARD		No	1884-1897	MI, Petosky		Cert.	company made wooden prod.
A. BEERS	BEERS, A.	Unusual	Yes		CT		Poss.	double flyer type 1 and type 3
ANDREW BEERS	BEERS, ANDREW	Unusual	Yes					double flyer type 3 -Cummer
	BEGGS, THOMAS&JAMES	Wool	No	1790's	NC, Fayette		Cert.	AWC
M.C. BEIMS	BEIMS, M.C.	Flax	Yes		ML		Poss.	
	BELL, JAMES	Flax	Yes		NY, Cobbleskill		Cert.	NYC in 1827, Buffalo 1833-40
J. BENSON L.I. ED	BENSON, J.	Flax	Yes		VT		Poss.	L. I. ED" also on it -(poss. Ulster Co. NY?)
	BERGERON, LOUIS ELOY	Unusual	Yes	1790's	FRANCE, Paris		Cert.	* Fancy table-top wheels
	BERRES, PETER	Flax	No	1901-1902	WI, West Bend		Cert.	SWS
J.B. 1856	BIRCHFIELD, JOHN	Flax	Yes	1856	OH, Worthington		Cert.	
D. BIRD	BIRD, D.	Unusual	Yes					SWS- double flyer
	BISHOP, LUCIUS B.	Patent	No	1874	CAN. Nova Scotia		Cert.	* like Hathorn's
	BISSELL, WILLIAM	?	No		NY, Albany		Cert.	turner in 1833
SILAS BLACK	BLACK, SILAS	Wool	Yes		NY		Prob.	barrel tension
?. BLACKBURN	BLACKBURN, ?	Flax	Yes	1850'S	OH		Poss.	
J. BLAIR	BLAIR, J.	Flax	Yes		OH		Poss.	
W. BLAYNEY	BLAYNEY, W.	Flax	Yes					
	BLODGETT, A.& CO.	?	No		NYSkaneateles		Cert.	made Minor's heads
	BODENHEIMER, WILLIAM		Yes	1850'S	OH, Lancaster		Cert.	Gunsmith- see also "MATLOCK"
	BOICOURT, SAMUEL	?	No		IN, Princetown		Cert.	worked 1817-?
T. BOND	BOND, T.	Wool	Yes					broomstick legs, horn collar
BOOK-1826	BOOK,	Flax	Yes	1826				SWS
N. BOOKER 189 8	BOOKER, N	Flax	Yes	1898??				
G. BOWDISH	BOWDISH, G.	Winder	Yes					6 armed
A. BOWE	BOWE, A.	Flax	Yes					SWS
	BOWEN, ETHAN	Patent	No	1830,Jul13	RI, Providence		Cert.	
D.BOWVER	BOWVER, D.	Flax	Yes		PA, Bucks Co.		Prob.	Patent lost,"Spinning wool & cotton"(also Pat. on 1826, Dec29)
	BOYDEN, SETH	Patent	No	1826,Dec7	MA, Foxbrough		Cert.	Patent lost. Z. BRADLEY?
J.BOYER	BOYER, J.	Flax	Yes					SWS
	BRADEEN, SAMUEL	?	No					Working 1829-? (SWS- Alfred, ME 1830)
	BRADISH, C.	Patent	No	1824,Dec31	NY, Lowville		Cert.	Patent lost "Spinning machine"- see HUNT
Z. BRADLEY	BRADLEY, ZALMON T.	Flax, Unusual	Yes					double flyer-accelerated
Z. BRADLY	BRADLY, Z.	Flax,	Yes		NY, Ithica		Poss.	poss. Z. BRADLEY?
	BRADSHAW, JOHN		No	1815	MD, Hagerstown		Cert.	AWC-John Rutter became partner in 1820
	BRAIN, ?	Patent	No	1866				Pat. #59210 IN 1866
SAMUEL BRANT	BRANT, SAMUEL	Bat's head	Yes		PA		Poss.	poss. Maker
	BREAM, CONRAD	?	No	1770'S	NC		Prob.	

SPINNING WHEEL MAKERS

	Name	Type	Signed	Date	Location	Cert.	Notes
	BREWSTER, GILBERT	Patent	No	1824,Feb27	CT, Norwich	Cert.	Patent lost, "wool spinning wh.", also 3/13/1824 Pat.
W.M. BRIDE	BRIDE, W.M.	Wool	Yes		PA/OH	Poss.	
B.B.	BRIGGS, BARNABAS	Wool	Yes	1794-1801	ME, SDL	Cert.	Shaker Trustee
	BRODERSON, INGWART	?	No	1860's	WI, Milwaukee	Prob.	
	BROWN, FIELDS &		No	1820's	NY, Seneca Falls	Cert.	chair & wh. maker in 1820'S
	BROWN, J.	Patent	No	1815	RI, Providence	Cert.	Pat. wheel in 1815
	BROWN, JOHN	Patent	No	1813,May12	RI, Providence	Cert.	Patent missing "Spinning machine" also 2/12/1814, 7/11/1821, 5/20/1
L. BROWN	BROWN, L.	Unusual	Yes		NY	Prob.	Farnham type
S. BROWN	BROWN, S.	Unusual	Yes		NY	Prob.	Farnham style
T.D. BROWN	BROWN, TIMOTHY DEWEY	Patent type	Yes	1870-1882	CA, Oakland	Cert.	* SWS-"track" type
W. BROWN	BROWN, W.	Flax,	Yes		OH	Poss.	Ex. wh. supports
JEDEDIAH BROWNING	BROWNING, JEDEDIAH	Unusual	Yes	1820's	CT, Woodstock	Cert.	ATHM has unmarked one
	BROWN'S VERTICAL SPINNER	Patent	No	1825	NY, Kingsbury	Cert.	J.R. & J.B. Wheeller patentees
J. BRUCE	BRUCE, J.	Patent	Yes	1873	CAN.	Cert.	* J. BRYCE Canadian version
J.G. BRUNN	BRUNN, J.G.	Wool	Yes		PA	Poss.	SWS
	BRUSHNELL, WM.	Patent	No	1821,Aug30	VA, Harrisville	Cert.	Patent lost, "Spinning wool & cotton" See ALTOFFER
	BRYAN, CHARLES		No	1838-1857	PA, Plumstead Twp.	Cert.	
J. BRYCE	BRYCE, JOHN	Patent	Yes	1872	MI, Grand Haven	Cert.	* Pat. # 131657
E. BULLOCK	BULLOCK, EZEKIEL	Flax	Yes	1820	PA, Carlisle	Cert.	(Antiques Mag. Article April, 2001) in CARLISLE 1819-1844
	BULLOCK, WILLIAM	?	No		IN/PA	Cert.	AWC-1839-Lafayette,IN; 1820'S -Carlisle,PA; 1837-Shippensburg, F
	BURKETT, THOMAS	?	No	1770's	NC	Prob.	Esposito
J.W. BURKHART	BURKHART, J.W.	Patent	Yes	1868	MO, Camron	Cert.	* Pat.#81594, see "HAWKINS"
JOHN BURLEY PENN 1845	BURLEY, JOHN	Unusual	Yes	1845	PA, Fayette Co.	Cert.	stamped "1845 No. 13" double wheel & treadles, snake foot leg
R. BURROWS	BURROWS, R.	Flax	Yes		PA	Poss.	"R" is questionable
	BUSHWELL, JOHN	?	No	1820's	NH, Andover	Cert.	
BUTLER	BUTLER	Flax	Yes		NY		
I. BUTLER	BUTLER, IRA?	Wool	Yes		NY	Poss.	Barrel tension
	BUTNER, DAVID	?	No	1790's	NC	Prob.	
	BYRKIT, JESSE	Patent	?	1865	WI, Fairfield	Cert.	Pat.# 50094, gun maker?
	BYRKIT, JESSE	Wool	No	1865	IO	Cert.	Pat.#50094 sons made rifles
D.C.	C, D.	Flax	Yes		CT	Poss.	
D.R.C.	C, D.R.	Flax	Yes		NH/ME	Prob.	Shaker like
H.C.	C, H.		Yes		Scotland?	Poss.	"C" could be "G"
I.C.	C, I.	Flax	Yes				Shaker like, heavy maidens, stars by mark, single vase leg
J.C.	C, J.	Flax	Yes	1740	RI/CT	Poss.	dated 1740
J.C.	C, J.	Wool	Yes		NH	Poss.	sq. legs, sliding shaft tension
L.C.	C, L.	Flax	Yes				SWS
O.C.	C, O.	Wool	Yes				sliding table tension
R.C.	C, R.	Flax	Yes	1833			picked in, "1833 NO. 94"
R.C.	C, R.	Flax,Wool	Yes		PA	Poss.	
T.C.	C, T.	Unusual	Yes				double flyer type 2 (T-top) back post is board, hooks hold flyer
I. CADAN	CADAN, I	Flax	Yes		European?		I" poss. "L", "LINEN BOARD"on side

SPINNING WHEEL MAKERS

	Name		Type	Signed	Date	State/Loc	Cert.	Notes
C. CAIN	CAIN, C.		Flax	Yes		NY/PA	Poss.	signed on side of table
MA.TT CALHOON	CALHOON, MA.TT		Flax, Unusual	Yes		OH	Poss.	* Several seen
	CALVERT, WM. W.		Patent	No	1832,Mar31	MA, Lowell	Cert.	Patent lost,"Spinning wool", see Messenger & Southwick
	CAMPBELL, ASA		Patent	No	1834,Jun19	PA, Kingston	Cert.	Patent lost,"Spinning wool"
S.S. CAMPBELL	CAMPBELL, SIDNEY S.		Miner's heads	Yes	1820-1860	NH, Chesterfield	Cert.	produced heads
H. CARLON	CARLON, H.		Flax	Yes		PA?	Poss.	N" is backwards
T. CARLON	CARLON, T.		Flax	Yes		PA	Prob.	N" is backwards
M. CARPENTER	CARPENTER, M.		Flax	Yes		PA, Lancaster Co..	Cert.	seen by Bill Leinbach
A. CARR	CARR, A.		Flax	Yes	1780-1790's	PA,CHESTER Co.	Cert.	AWC
I. CARR	CARR, I.		Flax	Yes		OH/PA/MD	Poss.	wheel post supports and a "T" strocher, thick table
W. CARR	CARR, W.		Flax	Yes				
T. CARTER	CARTER, T.		Flax	Yes		NY	Prob.	*
ALFRED CHAMBERLAIN	CHAMBERLAIN, ALFRED		Minor's heads	No	1820's	NH, Chesterfield	Cert.	produced heads
ALLISON CHAMBERLAIN	CHAMBERLAIN, ALLISON		Minor's heads	No	1820's	NH, Chesterfield	Prob.	produced heads
	CHAPIN, ELIPHAT(E)			No		CT	Prob.	poss.. wh. maker
S. CHENEY	CHENEY, SILAS E.		Flax	Yes	1776-1821	CT, Litchfield	Cert.	ad in Litchfield Monitor 9/29/02
	CHILD, JONAS		?	No	1805	NH, Packerville	Cert.	AWC-after 1805 made chairs/wheels; also in Gardnier, MA IN 1827
CHIPMAN	CHIPMAN			Yes				
A+J CHURCH	CHURCH, A+J		Unusual	Yes		CT	Prob.	chair wheel
	CHURCH, ELI			No	1820's	NY, Homer	Cert.	Esposito.
	CHURCH, J.		Patent	No	1836		Cert.	Pat.# 9669, no info
	CHURCH, JAMES JR.		Patent	No	1830,May22	CT, Hartford	Cert.	Patent lost,"Spinning machine"
	CHURCH, JOSHUA G.		Patent	No	1880	TX, Breckenridge	Cert.	Pat.# 229093
	CHURCH, UZZIEL		Flax, Reels	No	1820's	IN, Brownville	Cert.	see:"CHENEY SPEARS", quillers too
	CHURCH, WILLIAM		Patent	No	1827,Jun11	ENG., Birmingham	Cert.	Patent lost,"Spinning wool & cotton"
D. CLARK	CLARK, D.		Flax	Yes		PA	Poss.	
	CLARK, S.W.		Patent	No	1868	W, (Prairie Du Chien	Cert.	Pat.# 71047 pendulum type
	CLARKE, WILLIAM		Patent	No	1830,Oct1	NY, Poultny	Cert.	Patent lost,"Spinning machine"
T. CLIFF + P	CLIFF, T.			Yes		VT, Shelburne?	Poss.	Esposito.
	CLIMER, LUDWIG			No	1870'S	NC	Prob.	working 1870'S
	COCHRAN, JAMES JR		Patent	No	1872	CAN, Nova Scotia	Cert.	Pat.# 129459, "table top spindle"
	COFFIN, PETER			No	1772-1859	ME	Poss.	poss. Maker
	COIR, A.			No	1770's	NY	Poss.	
W. COLDER	COLDER, W.			Yes		PA	Poss.	
	COLE, CHARLES L.		Patent	No	1868	MI, Richmond	Cert.	Pat.# 84261 11/24/1868
M. COLLINS	COLLINS, M.		Wool	Yes		ME/NH	Poss.	Shaker like
	CONE, OLIVER			No	1795	MA, Greenfield	Cert.	AWF- his wheels sold by Wise Grinnell (also a wheel maker)
	CONRADT, GEO. M.		Patent	No	1834,Mar10	MD, Fredericktown	Cert.	Patent lost,"Spinning wool",(see SYKES)
JOHN COOK	COOK, JOHN		Unusual	Yes		NY	Poss.	solid double wheel
	COOK, NATHAN TOPPING		Flax	No	1792-1822	NY, Bridgehampton	Cert.	SWS-"Suffock Co."
	COPELAND, A.		Patent	No	1824,Dec24	PA, Phila.	Cert.	Patent lost "Spinning machine", see POOL
N. CORDRAY	CORDRAY, N.		Flax	Yes		OH	Poss.	

SPINNING WHEEL MAKERS

	Maker						Notes
M.T. CORLIES 1823 No 23	CORLIES, M.T.		Flax	Yes	1823	PA	Poss.	parlor style, dated 1823 NO. 23, 6 piece drive wheel rim
	CORNY, V.			No				
	CORRILL, J.		Patent	No	1826,Jun15	OH, Harpersfield	Cert.	Patent lost,"Spinning wool"see Rogers,WM. P.")
	COTTON, JAMES			No				Ulrich book
	COURIER		Miner's heads	No	1882	NH, Chesterfield	Cert.	bought out Pierce factory in 1882
D. COVE	COVE, D.		Quiller	Yes				double wheel quill winder
	COX, WILLIAM			No	1760's	PA, Phila.	Cert.	AWF-apprentice to John Shearman, wheelmaker
	CRAIG, ROBERT			No	1757-1777	MA, Leichester	Cert.	AWF-poss. "CRAGE", made reels and quillers & did repairs
C.M. CRANDALL	CRANDALL, CHARLES M.		Unusual	Yes	1875-1886	PA, Montrose	Cert.	made toys,flax,wool,winders,heads
JACOB CRANE	CRANE, JACOB		Flax	Yes	1812?	NY, Long Island	Poss.	poss."JACOB GRANT"
JOHN CRANE	CRANE, JOHN		Flax	Yes		VT	Poss.	
B. CREEN	CREEN, B.		Flax, Wool	Yes		PA/NJ	Poss.	*
	CRICHFIELD, ARTHUR		Patent	No	1828,Apr24	OH, Union Twp.	Cert.	Patent lost, "Spinning wool & cotton"
	CROSBY, JAMES			No	1814	DE, Wilmington	Cert.	AWF
	CRUNK, JAMES			No	1790's	NC	Prob.	poss .maker
	CRUNK, MICHAEL			No	1790's	NC	Prob.	poss. Maker
	CURRANT, DAVID		Patent	No	1850	KY, Crittendom	Cert.	Pat.# 7614 9/3/1850
J.C.	CURRY, JAMES		Flax	Yes	1812	PA, Milan, Bradford	Cert.	paper on underside of table, info about wheel and maker
	CURRY, JOHN			No	1803-1818	KY, Paris	Cert.	AWC-wheelwright
	CURTIS, ANSON			No	1800 early	NY	Poss.	SWS
H. CURTIS	CURTIS, H.		Wool	Yes		CT	Poss.	3 other initial
T.C.	CUSHMAN, THOMAS		Flax, Wool	Yes	1758-1816	ME, Alfred	Cert.	* Shaker Trustee (1801-1809)
	CUTTER, JOHN		Wool	No	1850	IN, Carrol Co.	Cert.	adv. wool wheels for $4.25
	CUTTER, JOHN JR.		Wool	No	1850	IN, Carrol Co.	Cert.	adv. wool wheels for $4.25
D.	D		Flax	Yes		MA	Poss.	
A.D.	D, A.		Flax	Yes		LA	Poss.	
E.D.	D, E.		Unusual	Yes		PA	Poss.	double flyer type 1
J.B.D.	D, J.B.		Flax	Yes		CAN.	Prob.	ornate metal treadle
L.D.	D, L.		Flax, Wool	Yes				
L.O. M.D.	D, M.		Flax	Yes				12 spoke,zig zag decoration,see "L.O."
W. DANIELS	DANIELS, W.		Flax	Yes		PA	Prob.	wheels and swifts, prob. Daniel's son
A. DANNER	DANNER, AARON		Flax	Yes	1821-1864	PA, Manheim	Cert.	* Irish castle,sax. ,5 arm winder (lived 1803-1881)
DANIEL DANNER	DANNER, DANIEL		Wool	Yes		PA/OH	Poss.	found at N. Union Shaker villaage, acorn feet
B. DAVIS	DAVIS, B		Flax	Yes		PA, Pittsburgh	Cert.	
W. DAVIS PITTSBURGH	DAVIS, W.		Wool	Yes				
W.H. DEAN	DEAN, W.H.		Miner's heads	Yes		NY, Skaneateles	Cert.	E-Bay auction has "_____ONT" (CANADA) ON IT also.
H. DELAND	DELAND, H.		Patent	Yes	1866	CAN, Strathroy,ONT.	Cert.	On "LORING" wh.
S. DELL	DELL, SOLOMON		Patent	No	1866	CAN, Adelade	Cert.	* Can. Pat., lever action wh. b.1823, d.1899
	DELL, WILLIAM		Miner's heads	No				SWS
D. DEMMING & PIERCE CO	DEMMING, D.		Miner's heads	No		NY, Onondaga Co.	Cert.	in business with A. Miner
	DEMMING, S.							in business with A. Miner
J. DERBY	DERBY, J.							

SPINNING WHEEL MAKERS

		Patent(CAN)			Cert.	
DEXTER	DESMARAIS, PIERRE			MA, Holyoke		SWS
	DEXTER, ?	Wool	Yes	KY	Poss.	
	DIENMAN, CONRAD	Wool	No	OH, Zoar	Cert.	prob. wool German immigrant marked wheels with his name
	DIETZ, A.F.		No	NY, Scholarie	Prob.	poss. maker 1880'S
	DOBSON, HENRY B.		No	NC	Poss.	poss. maker 1790'S
	DOMINY, FELIX	Flax	No	NY, East Hampton	Cert.	SWS-"Suffock Co."
	DOMINY, NATHANIAL IV	Flax	No	NY, East Hampton	Cert.	SWS-"Suffock Co."
	DOMINY, NATHANIAL V	Flax	No	NY, East Hampton	Cert.	SWS-"Suffock Co."
JOSEPH DOUGHTY	DOUGHTY, JOSEPH	Unusual	Yes	ENG. York	Cert.	* auto-winding bobbin device on parlor wheel
	DOUGLAS, SAMUEL		No	CT, New Haven	Cert.	AWC-had wh. factories with Wait Garret and Samuel Douglas Jr.
	DOUGLAS, SAMUEL JR.		No	CT, New Haven	Cert.	AWC-had Spinning wh. factories with father Samuel Sr.
C. DOUGLASS	DOUGLASS, C.	Wool	Yes	ME.	Poss.	
DOW	DOW	Winder	Yes	CT.	Prob.	* saxony based 6-arm winder
OLIVER DOW J	DOW, OLIVER(JR.?)	Winder	Yes	NH	Poss.	June 15, 1789" on winder
	DOWNING, SAMUEL		No	NH, Deering	Prob.	SWS
N.D.	DRAPER, NATHANIEL	Wool	Yes	NH, Enfield	Cert.	* Shaker Trustee 1795-1821
	DRISSEL, JOHN		No	PA, Bedminster	Prob.	SWS-"Bucks Co."
D. DRYER	DRYER, D.	Flax, Wool	Yes	NY	Prob.	single nut barrel tension
JAMES DUFF	DUFF, JAMES	Flax	Yes			poss. Scotch
	DUNSHEE, W.		No		Poss.	Esposito
H.E. DYER	DYER, H.E.	Winder	Yes	NY		4 arm winder
N.E.	E, N.	Flax	Yes			
	EASTMAN, ENOCH	Flax, Wool	No	NH, Andover	Prob.	
	EASTMAN, VERSAL		No	NH, Andover	Prob.	poss...maker
	EATON, JOHN		No	NH, Gaffary	Prob.	John Eaton Manufactory-1770'S
	EDEN, JOSHUA	Wool?	No	SC, Charleston	Cert.	AWC & AWF
J. EDGERLY CANTERBURY	EDGERLY, JOSIAH	Flax	Yes	NH, Canterbury	Cert.	* Shaker Trustee," CANTERBURY" on wheel
W.J. EDKIN 1847-8	EDKIN, W.J.	Wool	Yes	PA	Poss.	SWS
T. J. EDKINS 1848	EDKINS, T.J.	Flax	Yes	PA	Poss.	saxony-not traditional style -poss.from farther east?
ELEVEGOOD	ELEVEGOOD,	Flax	Yes			SWS
	EMERSON, JOSEPH		No	ME, Falmouth	Cert.	AWF-1740'S and before
	EMERSON, WILLIAM G.		No	NH, Canaan	Prob.	SWS
	EMORY, JESSE		No	NH, Weare	Prob.	SWS
	ENGEL, KOLOMAN	Patent	No	Austria-Hungary	Cert.	Pat.# 580452
	ESH, JOSEPH	Flax	No	WI, Beaver dam	Cert.	SWS
A.F.	F, A.	Head	No	NJ	Poss.	SWS
C.F. A HLS	F, C		Yes			SWS
D.F. 1847	F, D.	Flax	Yes			SWS
G.B.F.	F, G.B.	Flax	Yes	CT	Poss.	had attached winder
I.Fy	F, I.'	Flax	Yes			see "I.F.Y."
I.F.	F, I.	Wool	Yes			
I.F.	F, I.	Wool	Yes	PA	Prob.	

SPINNING WHEEL MAKERS

Mark	Name	Type	Signed	Date	Location	State/Area	Cert.	Notes
S.F.	F, S.		Flax	Yes	1839		CT, Pleasant Valley	Poss.	"S" reversed
T.F.	F, T.		Flax	Yes					
W.F.	F, W.		Flax	Yes					
	FAIRBANKS, HENRY		Patent	Yes	1866		IL, Adams Co.	Cert.	SWS Pat.# 58015, see "WILSON"
	FALTZ, ANDRESS		Flax	Yes	1801-1834		NC, Old Salem	Cert.	4 wh. dated 1801-1834
W. FANCHER	FANCHER, W.		Unusual, Wool	Yes			NY	Poss.	*
FARNHAM AND LEWIS	FARNHAM,		Wool	Yes			NY, Owego	Cert.	Tioga Cty. Hist. Soc. has wheel
	FARNHAM, CHARLES		Flax	No	1830?		NY, Owego	Cert.	SWS
F.A. FARNHAM OWEGO	FARNHAM, FREDERICK AUGUSTUS		Wool, heads,	Yes	1818-1887		NY, Owego	Cert.	* son of Joel Farnham. working 1840 on
J. FARNHAM OWEGO	FARNHAM, JOEL		Flax, Wool	Yes	1774-1858		NY, Owego	Cert.	* flax,wool,Pat.,heads,winder, wheels started about 1794
J. FARNHAM JR. NEAR	FARNHAM, JOEL JR.		Unusual	Yes			NY, Owego	Cert.	
	FARNHAM,JOEL		Patent	No	1825,Jun28		NY, Tioga	Cert.	Patent lost,"Spinning wool & cotton"(see WITTSE)
J. FAUNCE	FAUNCE, J.		Flax	Yes				Cert.	
FRANK FELL	FELL, FRANK		Flax	Yes			WI, Mayville	Cert.	late 1800's ads
	FELL, FRANK		Flax	Yes	1865-1935		WI, Mayville	Cert.	SWS
	FELTCH, ABIJAH			No	1828				Esposito.
I. FERGUSON	FERGUSON, I.		Flax	Yes			NH, Weare	Prob.	
	FETTER, JACOB SR.			No	1777 prior		PA, Lancaster	Cert.	AWF-died in 1777
RUSSEL FIELD	FIELD, RUSSEL		Unusual	Yes			CT	Prob.	double flyer type 1
	FIELDS & BROWN			No	1820's		NY, Seneca Falls	Cert.	chair & wh. maker in 1820's
	FIESTER, JACOB			No	1820's		PA, Columbia Co.	Cert.	chair & wh. maker in 1820's
	FITZGERALD, THOMAS			No	1760		PA	Cert.	run away bond servant ad in Phila. paper
	FIZER, SAMUEL			No	1820		VA, Fincastle	Cert.	AWC-seeking wheelwright for spinning wh. Shop
R. FLEMING	FLEMING, R.		Flax	Yes			VT	Poss.	
T. FLEMING	FLEMING, T.		Flax	Yes					sq. spokes chanfered
	FLOOD, J.M.		Patent	No	1865		MO, Fulton	Cert.	Pat.# 51532
	FORD, SIMON			No	1847		WI, Watertown	Cert.	SWS
	FOUST, J. WILSON		Patent	No	1868		PA, Evanburg	Cert.	Pat.# 83970, see "JOHNSON"
F.A. FOX	FOX, FREDERICK A.		Flax	Yes	1836		PA, Morgantown	Prob.	SWS-"Berks Co."
	FOX, GEORGE			No	1771		PA, Phila.	Cert.	turner & wh. maker in 1771
IA: FOX	FOX, IA (JACOB)		Flax	Yes	1808-1862		PA, Berks Co.	Prob.	*
J. FOX	FOX, JACOB			Yes	1839 after		PA	Prob.	poss. JACOB
	FOX, JAMES			No	1800's?		PA, West Bradford	Prob.	SWS
	FOX, WILDER M.		Patent	No	1886		ME,Upper Stillwater	Cert.	Pat.# 339981
N.F. ALD	FREEMAN, NATHAN		Wool	Yes	1822-1862		ME, Alfred	Cert.	* Shaker Trustee
E. FROST 1836	FROST, E.		Winder	Yes	1836		MA	Poss.	* 4 armed,"made round wooden boxes too"
FRYSON	FRYSON		Flax	Yes			PA	Poss.	1st letter not cert., poss. further east
?. FULTON	FULTON, ?		Flax	Yes					
	FURR, HENRY			No	1770's		NC		working in 1770'S
I.FY	FY,I.		Flax	Yes			OH	Prob.	extra wheel supports
B.G.	G, B.		Flax	Yes			PA	Poss.	16 spokes
D.G.	G, D.		Wool	Yes					SWS

SPINNING WHEEL MAKERS

E.G.	G, E.		Flax	Yes			Poss.	
H.G.	G, H.		Flax	Yes	1767			G" could be "C", "1767" on it
JDG	G, JD		Flax					
N.G. NO. 5 60	G, N.		Flax, Unusual	Yes		VA/MD	Poss.	* No. 554 is like "MATT CALHOON" wheel, No. 560 is a saxony
S.G.	G, S.		Flax	Yes				SWS
	GAGE, CALEB STRONG			No	1830-1840's	MA, Essex	Poss.	AWC- poss. chair wheel maker, made chairs and wheels
	GAINES, JAMES P.		Patent	No	1882	KY, Kuttawa	Cert.	SWS
WAIT, GALLUP AND W-----	GALLUP, HENRY		Flax, Patent	Yes	1868	WI, Watertown	Cert.	SWS
G. GANDER	GANDER, G.		Wool	Yes		PA, Bucks Co.	Prob.	SWS-"small""
	GARDNER,		Wool	No				SWS
	GARDNER, THOMAS			No	1789	NJ, Bottle Hill	Cert.	working IN 1789
	GARRET, WAIT			No	1810	CT, New Haven	Cert.	AWC-had Spinning wh. factories with Samuel Douglas
I. GERIS	GERIS, I.		Flax	Yes				SWS
	GIBSON, ALEXANDER			No	1819	OH, Cincinnati	Prob.	SWS
	GIBSON, JAMES			No	1780's	NC	Prob.	working in 1780'S
	GIESY, NICHOLAS		Unusual	No	1855-1877	OR, Aurora	Prob.	SWS-"lever action"
J. GILFORD	GILFORD, J			No				SWS
	GLOVER, BOYD		Patent	No	1866	Il., Camp Point	Cert.	Pat.# 56600, see "ODELL"
	GOODRICH, E.			No	1830's	ME	Poss.	SWS
F.A. GOODRICH	GOODRICH, F.A.		Hackle	Yes		MA, Hancock	Cert.	Shaker wh. maker ?
	GORMAN, JAMES			No		NH, Derryfield	Poss.	SWS
	GRAHAM, JOHN			No	1817-1830	NC	Prob.	turner working 1817-1830
JACOB GRANT	GRANT, JACOB		Flax	Yes	1812?	NY, Long Island	Poss.	poss. "CRANE", date poss.
S.B. GRAVES	GRAVES, S.B.		Wool	Yes		PA	Poss.	stamped twice on top of table
B. GREEN	GREEN, B.		Flax	Yes		PA	Poss.	
H. GREEN	GREEN, H.		Unusual	Yes		PA/NJ	Poss.	wool,Irish castle
J. GREEN	GREEN, JOHN		Patent	Yes	1866	IL, Joliet	Cert.	* Pat.# 59210
I. GREGG	GREGG, I.		Flax	Yes				poss. "I" is Joseph (ATHM)
J. GREGG	GREGG, J.		Winder	Yes	1836			1836" on 6 arm winder
JAMES GREGG	GREGG, JAMES		Flax	Yes				
JOSEPH GREGG	GREGG, JOSEPH		Flax	Yes				
	GREGG, W.			No		NH, Londonderry	Cert.	
GREGGS	GREGGS		Wool	Yes		PA, Berks Co..	Poss.	SWS
G. GREIDER	GREIDER, G.		Flax	Yes		PA	Poss.	reel also
J. GREIDER	GREIDER, J.		Flax	Yes		PA	Poss.	reel also
S.R. GRIVES	GRIEVES, S.R.		Wool	Yes		MI/OH	Poss.	poss. misreading of "S.B.GRAVES"
	GRIGGS, JOHN JR.		Patent	No	1824,Dec30	CT, Ashford	Cert.	Patent lost,"Spinning wool"
	GRINNELL, WISE			No	1795	MA, Greenfield	Cert.	AWF- made wheels and also sold some made by Oliver Cone
A.H. 1816	H, A.		Flax, Wool	Yes	1816-1827			wool-"1816", flax-"1827"
B.H.	H, B.		Flax, Wool	Yes		CT	Poss.	
C.H.	H, C.		Wool	No		NC, Old Salem	Prob.	

SPINNING WHEEL MAKERS

Name	Code	Type	Yes/No	Date	Location	Cert.	Notes
D.H.	H, D.		Yes				SWS-textile tool
E.H.	H, E.	Flax	Yes			Poss.	
H.H. 1812	H, H.	Flax	Yes	1812	PA	Poss.	
I.H.	H, I.	Flax	Yes	1831-1838	PA/OH	Cert.	1831,1832,1838, dates seen
I.H. 1827	H, I.	Flax	Yes	1827	PA	Poss.	* picked in, Aspundth
J.H.	H, J.	Flax	Yes		MA	Poss.	
J.H.	H, J.	Wool	Yes		PA, Haverford Twp.	Prob.	SWS
J.N.H.	H, J.N.	Flax	Yes		PA	Poss.	seen on e-Bay, poss. European
M.H. 1803	H, M.	Flax	Yes	1803	PA	Poss.	
M.T.H.	H, M.T.	Flax, Wool	Yes		VA	Poss.	small wool wheel
M.V.H. 1838	H, MV	Flax	Yes	1838		Poss.	1838
O.H.	H, O.	Flax	Yes		CT	Poss.	poss. OBEDIA HIGINBOTTOM
O.H.	H, O.	Flax	Yes		PA	Poss.	3 seen in PA.
H.O.H.	H, OH	Reels	Yes	1800-1806	PA	Poss.	reels dated "1800"&"1806"
P.H.	H, P.	Winder	Yes		ME	Poss.	4 arm winder
S.H.	H, S.	Flax	Yes				
U.H.	H, U.	Flax	Yes			Cert.	Shaker like -plain
O. Hagen	HAAS, JOHANNES	Winders	No	1814-1856	PA, Schulkill Co.	Cert.	A turner and yarn reel maker
	HAGEN, O.	Flax	Yes			Cert.	"a" is reversed, poss. Eur.
P. HALL	HALL, JOHN		No	1691-1760	PA, Chester Co.	Poss.	Chester Co. furn. makers
P. HALL	HALL, P.	Flax	Yes		DE, Dover	Poss.	P" may be PURNELL
D. HANEY	HANEY, D.	Wool	Yes				sliding table tension, very plain, e-bay pictures
HARMONY	HARMONY	Wool	Yes	1820's	PA, Butler Co.	Poss.	Esposito- poss. Maker
	HARRIS, NATHANIEL	Patent	No	1829,Jan21	OH, Fairfield Co.	Cert.	Patent lost,"Spinning wool & cotton by hand"
C. HARRISON	HARRISON, C.	Flax	Yes		PA	Poss.	
W.H. HART	HART, W.H.	Flax	Yes		PA	Poss.	
	HARTMAN, HENRY		No	1863	WI, MILWAUKEE	Prob.	SWS
	HARTWELL, ISAAC B.	Patent	No	1839,Jun7	VT, Northfield	Cert.	Patent lost,"Spinning wool"
HATCH	HATCH, JOEL		No	1812	NY, Shelburn	Prob.	Esposito, see Cummer Pg.152 poss...N.H.?
	HATCH, WILLIAM P.	Patent	No	1871	ME, Lincoln Centre	Cert.	Pat.# 115465
G.H. HATHORN	HATHORN, GEORGE H.	Patent	Yes	1871	ME, Bangor	Cert.	* Pat.# 121517
	HAVEN, GIDEON		No		MA, Framingham	Cert.	son-in-law of JEREMIAH PIKE-maker
HAWKINS BROTHERS	HAWKINS	Patent	Yes	1868			made & dist. Burkhart wh.
C. HEESE	HEESE, C.	Flax	Yes				SWS
HEIM 1802	HEIM		Yes	1802	PA	Poss.	
	HEIZMAN, PHILIP	Flax	No	1814-1896	WI, Watertown	Prob.	SWS
A. HEMPHILL	HEMPHILL, ANDREW	Wool	Yes	1772-1844	NY, Troy	Cert.	horn collar, SWS gave dates and location
W. HENDERSON	HENDERSON, W.	Flax	Yes		OH	Poss.	N is backwards
	HENNEN, JOHN		No	1754	PA	Cert.	run away bond servant ad in Phila. paper
	HENRY, JOHN		No		OH, Marietta	Prob.	SWS
S. HENRY 1812	HENRY, SAMUEL	Flax, Unusual,	Yes	1780-1816	PA, Lancaster	Cert.	Sax.-1812, Irish 4 leg-1806, wool and Windsor chairs
	HERRICK, EB.	Patent	No	1810,Aug17	NY, Albany	Cert.	Patent lost,"Spinning machine, domestic"

SPINNING WHEEL MAKERS

	Name	Type		Date	Location		Notes
	HIBBARD, JOSEPH		No	1700-1737	PA, Chester CO.	Cert.	Chester Co. furn. makers
	HICOX		No	1830's	MI, Northville	Cert.	wh. maker IN 1830'S
	HICOX, "FATHER"		No	1830's	MI, Northville	Cert.	
OBEDIAH HIGINBOTHAM	HIGINBOTHAM, OBEDIAH	Flax	Yes	1750-1804	CT, Pomfret	Cert.	
	HILL, JOHN		No	1790's	NC	Prob.	working in 1790'S
S. HILLARD	HILLARD, S.	Flax, Wool	Yes				unusually lg. wh. on flax wh.
HISD 1821	HISD	Flax	Yes		OH/PA	Poss.	picked in top of table
P. HOFFMAN	HOFFMAN, P.	Flax	Yes		PA	Poss.	stamped on top of table
	HOFFMAN, PHILLIP		No	1842	PA, Chester Co.	Cert.	Chester Co. furn. makers-census:"turner & wheelmaker"
J.L. HOLCOMBE	HOLCOMBE, J.L.	Winder	Yes		PA, Wortlesburgs	Cert.	Quill winder, town & state on it
	HOLLAND, THOMAS		No	1800's	MA	Prob.	poss. wh. Maker
J.H.	HOLMES, JOHN (ET AL)	Flax, Wool	Yes	1801-1856	ME, SDL	Cert.	* Shaker trustees used this mark 1801-1856
	HOLMES, OLIVER	Flax	Yes	1776-1841	ME, SDL	Cert.	Shaker, like SRAL, made in 1793
J. HOLMS	HOLMS, J.	Flax, Wool	Yes		ME	Poss.	very Shaker like, stamped on top of table
I. HOMSHER	HOMSHER, I.	Flax	Yes		PA	Poss.	* very large bobbin and flyer unit
S. HOOD	HOOD, S.	Wool	Yes				narrow rim, direct drive
I. HOOK	HOOK, I.	Flax	Yes		PA, Lancaster	Prob.	T-stretcher
J.L. HOOK	HOOK, J.L.	Flax	Yes		PA, Lancaster	Prob.	T-stretcher
	HOOPES, ABRAHAM		No	1823 Died	PA, Chester Co.	Cert.	Chester Co. furn. Makers-apprentice of Sampson Barnet
D. HOOVER	HOOVER, D.	Flax	Yes				"H" stamp looks like a "N" (one found in MD)
	HOPKINS & HOWE	Miner's heads	Yes		NH, Chesterfield	Cert.	poss. error of wheel head makers
	HOPKINS, BENJAMIN PIERCE	Miner's heads	Yes		NH, Keene	Cert.	made Miner's heads
	HOPKINS, E.	Miner's heads	No	1850's	NH, Chesterfield	Cert.	made Miner's heads
H.W. HOPKINS	HOPKINS, H.W.	Wool	Yes	1824	CT, Litchfield	Prob.	date in pencil on table table
J. HOPKINS	HOPKINS, J	Unusual	Yes				Double flyer type 2, (T-top)
JOHN HOPKINS	HOPKINS, JOHN	Wool	Yes				
	HOPKINS, JONATHON	Miner's heads	No		NH, Chesterfield	Prob.	prob. head maker
	HOPKINS, JOSEPH		No	1806	OH, Chillicothe	Cert.	ad in Scioto Gaz. 1806
R. HOPKINS	HOPKINS, R.(RICHARD?/JR.?)	Miner's head	Yes		NH, Chesterfield	Cert.	
	HOPKINS, RICHARD HENRY	Miner's head	No	1831-1877	NH, Chesterfield	Prob.	prob. head maker
	HOPKINS, RICHARD JR.		No	1820's	NH, Chesterfield	Prob.	working 1820'S
	HOPKINS, SAMUEL F.		No		NH, Chesterfield	Prob.	
W. HOPKINS	HOPKINS, W	Flax	Yes		CT, Litchfield	Poss.	SWS
W.W. HOPKINS	HOPKINS, W.W.	Wool	Yes				SWS
WILLIAM HOPKINS	HOPKINS, WILLIAM	Flax,Wool	Yes		CT, Litchfield	Prob.	
	HOPKINS, WILLIAM W.	Miner's heads	No		NH, Chesterfield	Prob.	
	HOTCHKISS, T.DWIGHT	Patent	No	1864	CT, Guilford	Cert.	Pat.# 42661 improved head
D. HOUGHTON	HOUGHTON, D.	Wool	Yes				
	HOWARD, JAMES		No	1806	OH, Chillicothe	Cert.	ad in Scioto Gaz. 1806
HOPKINS AND HOWE	HOWE, HORACE	Miner's head	Yes	1860-1870	NH, Chesterfield	Prob.	SWS
I. HOWELL	HOWELL, I.	Wool	Yes				
	HUDSON, EDWARD	Patent	No	1822	IN, Brookland	Cert.	ads in 1822 - making Pat. wh.

SPINNING WHEEL MAKERS

D.I. HUMAS	HUMAS, D.I.		Flax	Yes		PA, Birdsburo	Poss.	SWS-"I" may be an "H"
S. HUMES LANCASTER	HUMES, SAMUEL		Flax	Yes	1753-1836	PA, Lancaster	Cert.	* dates on wh. 1791-1831 (came from Ulster, Ireland)
	HUNT, H.		Patent	No	1824, Dec31	NY, Lowville	Cert.	Patent lost "Spinning machine" (See BRADISH)
	HUSE, WARREN.D.		Patent	No	1870	NH, Guilford	Cert.	Pat.# 108356 IN 1870, poss. Industrial
D.I.	I, D.		Flax	Yes		NY, Long Island	Poss.	
D.M.I.	I, DM		Flax	Yes				picked into table
P.IEN	IEN, P.		Wool	Yes				
	IPITTEL, FREDRICK		Flax	No	1867	WI, Oshkosh	Prob.	SWS
J.J. 1828	J, J.			Yes				SWS
F. JACKSON	JACKSON, F.		Flax	Yes				SWS
	JACKSON, JOSEPH & SONS		Wool	No	1865-1890's	TX, Lexington	Cert.	prob. wool wheels
I. JACOB	JACOB, I.		Flax, Winder	Yes		PA	Prob.	* also "J.JACOB" on quiller, "J.JACOB" saxony
J. JACOBS	JACOBS, J.		Winder	Yes		PA	Poss.	box type sm. wool wh.
	JACQUES, RICHARD			No	1775	NJ, New Brunswick	Cert.	chair & wh. maker in 1775
JASON 1873	JASON		Wool	Yes	1873	PA	Poss.	"1873", under sliding tension
T. JEDKIN 1848	JEDKIN, T.			Yes	1848	PA	Poss.	"1848"
	JOHNSON		Patent	No	1870			Pat.# 109116
A. IOHNSON	JOHNSON, A.		Flax, Wool	Yes		PA	Poss.	"I" is prob. "J"
J. JOHNSON	JOHNSON, J.		Flax	Yes		PA		SWS
	JOHNSON, JAMES L. & FOUST		Patent	No	1868	PA, Evansburg	Prob.	Pat.# 83970 "accordian" wh.
	JOHNSON, MOSES			No	1832	NY, Buffalo	Prob.	working 1832
	JOHNSON, THOMAS		Patent	No	1870	MI, Ruby	Cert.	Pat.# 109016
	JOHNSTON, JAMES			No	1855	NY, NYC	Cert.	turner in 1855
	JOHNSTON, JAMES			No	1770's	NC	Prob.	working in 1770'S
	JOHNSTON, THOMAS			No	1870's	MI, St. Clare	Prob.	working in 1870'S
A. JONES	JONES, A.		Flax	Yes		CT	Poss.	
IRA JONES	JONES, IRA		Tool	Yes				SWS
	JONES, ISAAC			No	1790's	NC	Poss.	poss. wh. Maker
	JONES, NOEL		Patent	No	1812, Mar14	NY, Madison Co.	Cert.	Patent lost, "Spinning flax, reeling & spooling"
IOSEPH	JOSEPH		Flax	Yes				
L. JUDSON	JUDSON, L.		Flax, Unusual	Yes		LA	Poss.	* double flyer type 1
HJULBACK PPS	JULBACK, H		Flax	Yes			Prob.	threaded rods supporting wheel supports
A.K.	K, A.		Unusual	Yes		VT, Weathersfield	Poss.	double flyer, flax, some AK whs.found in CT.
B.K.	K, B.		Flax	Yes		MA	Poss.	
G.K.	K, G.		Unusual	Yes				double flyer type 1
H.K. 1813 714	K, H.		Flax	Yes	1813	PA/OH	Poss.	picked in, found in MD.
I.K.	K, I.		Flax	Yes		PA/MD	Poss.	
I.K.	K, I.		Flax	Yes	1832	PA/OH	Poss.	one dated 1832
I.H.K.	K, I.H.		Flax	Yes		PA	Poss.	
I.H.K. 1817	K, I.H.H.		Wool	Yes	1817	PA, Birdsboro	Poss.	SWS
J.K.	K, J.		Unusual	Yes		CT	Poss.	double flyer type 1, diamond chip carving
JHK+JNK	K, JH		Wool	Yes				poss. father & son team?

SPINNING WHEEL MAKERS

N.K.	K, N.			Flax		Poss.	NK run together & picked in
T.K. 1799	K, T.		1799	Flax		Yes	found in IN.
T.K.	K, T.			Flax	PA	Yes	
T.F.K.	K, T.F.			Tool	VA	Yes	SWS
S. KASSON	KASSON, S.			Flax	CT	Yes	J.COLFIX" on side, dist.?
F. KEDEY	KEDEY, F.			Flax		Yes	
I. KEDEY	KEDEY, I.			Flax	PA	Yes	eastern PA.
I. H. KEIM	KEIM, I.H.			Wool	PA?	Yes	
	KEISLING, SAMUEL		1880	Patent	TN, Monroe	No	Pat.# 224700
B. KEITH	KEITH, B.			Flax		Yes	SWS
J. F. KEIZ	KEIZ, J.F.			Flax	OH (S.E.)	Yes	
W. KELIA	KELIA, W.			Flax	NH	Yes	*
	KELLER				PA, Phila.	No	poss. wh. Maker
	KENDALL, JOHN J.		1875	Patent	NC, Greensbourough	No	Pat.# 170630
S. KENNEDY	KENNEDY, S.						SWS
PETER KETCHEL	KETCHEL, PETER			Flax	NJ	Yes	SWS
W. KILBOURN	KILBOURN, W.			Unusual		Yes	double wheel
J. KILLIAN	KILLIAN, JOHN or JACOB		1838-1850	Flax	PA, Lancaster	Yes	* Jacob is father to John. John working in 1838-1850
P. KILLIAN	KILLIAN, P.			Flax	PA	Yes	
	KIMBALL, ASA		1772-1863	Wool	CT, Windham	No	made whs. on Pudding Hill
	KIMBALL, JOSEPH			Flax	NH, Canterbury	No	ads in paper, Shaker imitator
P. KINE	KINE, P.		1824,Jul3	Patent	NY, Lowville	Yes	Patent lost "Spinning machine"
	KING, S.JR.			Wool		No	
J. KIPP	KIPP, J.			Unusual		Yes	double wheels & treadles, legs splayed out, marked at top
T. KIRK	KIRK, T.			Flax	PA	Yes	"P.KLAUSER WARRWNTED PENNA"(SIC)
P. KLAUSER WARRWNTED	KLAUSER, P.		1840's?	Flax	PA	Yes	* poss. 1840'S
B. KLEIN	KLEIN, B.			Flax	PA	Yes	1704 on table bottom poss. # not date
D. KLEIN	KLEIN, D.			Flax	PA	Yes	
D.B. KLEIN	KLEIN, D.B.			Flax	PA, Montgomery Co.	Yes	* many seen, winder also, "J" may look like an "I"
J. KLEIN	KLEIN, J.			Flax	CT	Yes	pos error of "KLEIN"
B. KLINE	KLINE, B.		1866	Patent	IL, Camp Point	No	Pat.# 57336 8/21/66
	KOELLER, HERMAN		1742-1807	Flax, Wool	NC, Old Salem	Yes	Several seen
JOHANNES KRAUSE	KRAUSE, JOHANNES		1850's	Flax	MI, Presque Isle	Yes	
AUGUST KREKLOW	KREKLOW, AUGUST			Flax	PA, Berks Co.	Yes	SWS
D. KRIM	KRIM, D.			Flax,Wool,Reel	PA	Yes	* arms missing, short legs,10-150 count on box dial
D. KUNKEL PA	KUNKEL, D.			Unusual	CT	Yes	double flyer type 2 (T-top)
A.L.	L, A.		1806		PA	Yes	plain, "1806" on it
A.L. 1806	L, A.			Unusual	CT	Yes	* chair wheels, two seen
E.L.	L, E.			Flax	CAN	Yes	
F.L.	L, F.		1828	Flax	NJ/NY	Yes	found on Martha's Vineyard in 1940
G.L. 1828	L, G.		1847	Wool	OH	Yes	one dated
J.K.L. 1847	L, J.K.						

SPINNING WHEEL MAKERS

Mark	Name	Type	Signed	Date	Location	Cert.	Notes
M.L. 14. 25	L, M.	Flax	Yes		PA	Poss.	southern highlands
M.L. 1818	L, M.	Flax	Yes	1818&1865		Prob.	SWS
M O L	L, M.O.	Flax	Yes				
R.L. 1771	L, R.	Flax	Yes	1771	PA/OH	Poss.	strong Dutch influence
S+L	L, S.	Winder	Yes		NY, Long Island	Poss.	4- armed winder
S.L.	L, S.	Flax	Yes		PA	Poss.	
T.L.	L, T.	Flax	Yes				
1810 H.LA N.298	LA, H.	Flax	Yes	1810-1817			"1817 H.LA N.576" also
T.L.	LACKEY, THOMAS	Flax	Yes		VA, Rockbridge	Prob.	Bev.-12 spoke
	LAINE, JOHN		No	1800's	NH, Candia	Cert.	poss. wh. maker in 1800, SWS- "LANE"
W. LAING	LAING, W.	Winder	Yes		NY	Poss.	
	LANDES, RUDOLPH		No	1732-1802	PA, Bedminister Twp.	Prob.	SWS-"Bucks Co."
	LAPHAM, BENJAMIN	Patent	No	1827, Jun29	NY, Queensbury	Cert.	Patent lost, "Spinning wool"
	LARADINIS, JOHN	Flax, Unusual	No	1895-1904	WI, Rosiere	Prob.	SWS-"vertical and flax wheels"
J. LAY	LAY, J.	Flax	Yes				looks early
J. LEICHT	LEICHT, JOHANN SIMON	Flax	Yes	1780-1861	NC, Old Salem	Cert.	dates seen: 1842, 1854, 1857
PETER E. LEICHT	LEICHT, PETER E.	Wool	Yes	1862	NC, Old Salem	Cert.	made wh. in 1862
I. LEIGHT 1829	LEIGHT, I	Flax	Yes	1829	PA	Poss.	SWS
	LEMUN, JOSEPH		No	1806	OH, Chillcothe	Cert.	ads in Scioto Gaz.
E. LEONARD	LEONARD, E.	Flax	Yes				
H. LEONARD	LEONARD, H	Flax	Yes				
G. LEVEGOOD	LEVEGOOD, G.		Yes		PA	Poss.	
	LEWIS,	Wool	Yes		NY, Owego	Cert.	see "Farnham and LEWIS"
A. LEWIS	LEWIS, A.	Flax	Yes				SWS
E. LEWIS	LEWIS, E.	Wool	Yes		NY	Poss.	F.A.Farnham Miner's head seen on Lewis wheel
N. LEWIS	LEWIS, N.	Flax, Wool	Yes		VT/MA	Poss.	single post barrel tension
	LEWIS, NATHAN	Patent	No	1817,Mar.29	NY, Canandaigua	Cert.	Patent lost, "small Spinning wh."
J. L'HEUREAU	L'HEUREAU, J	Unusual	Yes	1880's	CAN. L'Acadie, Que.	Cert.	* Double wheel and treadle
B.L.	LIBBY, BENNETT	Wool	Yes		NH, Canterbury	Cert.	Shaker, unique lap joint
	LICHTENTHAELER, DAVID	Wool	No	1823-1824	NC, Old Salem	Cert.	David there 1823-1824
A. LIGHT NO.368 MAY 6,	LIGHT, A.	Flax	Yes	1816			T- Stretcher
	LINCOLN, THOMAS		No	1778-1851	IN, Gentryville	Poss.	SWS
LINDSEY READING	LINDSEY, N.	Flax	Yes		PA, Reading	Poss.	SWS-"N. LINDSEY" mark
LITTLE AND RALPH	LITTLE,	Flax	Yes	1865?	WI, Sheboygan Falls		SWS
SETH LOCKHART	LOCKHART, SETH	Flax	Yes		WV	Prob.	c.1860-1864
W.H. LOGAN	LOGAN, W.H.	Unusual	Yes		PA	Prob.	* painted & stenciled
LORING	LORING	Wool	Yes		NY	Prob.	Barrel tension, great style
	LOTZ, HENRY	Patent	No	1892	WI, Horicon	Cert.	Pat.# 467654
LOVE + WHITELOCK	LOVE,	Flax, Wool	Yes	1790	PA, Phila.	Prob.	made windsor chairs too
B. LOVE	LOVE, B.	Flax	Yes		PA	Prob.	seen by Bill Leinbach
J. LOVE	LOVE, J.	Flax	Yes		PA, Phila.	Poss.	
S. LOVE	LOVE, S.						

SPINNING WHEEL MAKERS

ALEXANDER LOW	LOVE, WILLIAM			No	1793-1806	PA, Phila.	Cert.	maker, see LOVE+WHITELOCK
W. LOWDEN	LOW, ALEXANDER	Flax	Yes		NJ, Freehold	Cert.	
	LOWDEN, W.	Wool	Yes		PA, Phila.		
	LYNCH, JAMES		No			Cert.	
A.A.M. NO.57 1865	M, A.A.	Flax	Yes	1857	PA	Poss.	on side of table
A.B.M.	M, A.B.	Flax	Yes		PA	Poss.	
D.M.	M, D.	Flax, Wool	Yes		ME	Poss.	poss. N.H., very plain
E.M. 1836	M, E.	Flax	Yes	1836	TN	Poss.	E.M. in script, poss. "E.W", T stretcher
F.R.M.	M, F.R.		Yes			Poss.	also has "C.WOLT" OR "WOGT"
G.M.	M, G.	Wool	Yes		PA/OH	Poss.	
H.M. 1844	M, H.	Wool	Yes	1844	PA	Poss.	4"s backwards
HM	M, H.	Flax	Yes			Poss.	H&M run together
I.M. 1811	M, I.	Flax	Yes	1811	TN	Poss.	
I.M.	M, I.	Flax	Yes		IN	Poss.	multiple stamp marks
I.A.M. 1843	M, I.A.	Flax	Yes	1825-1843	PA/OH	Poss.	dates seen:1825,1834,1843
J.M. 1841	M, J.	Flax	Yes	1841		Poss.	
L.G.M.	M, L.G.	Flax	Yes		NY	Poss.	solid wood double wheel - "L" not clear
N.M.	M, N.	Flax	Yes		MA	Poss.	N" reversed, raised stamp
N.M.	M, N.	Wool	Yes				
N.A.H.M.	M, N.A.H	Flax	Yes		LA	Cert.	several from Ursuline Convent in New Orleans
T.M.	M, T.	Unusual	Yes		CT, Danbury	Poss.	* double flyer type 2 (T-top)
T.M.	M, T.	Unusual	Yes				"Am. castle" style
T.H.M.	M, T.H.	Flax	Yes		LA	Poss.	2 stamps seen with differnet sixe letters
W.M.	M.W.	Flax	Yes				
W+M 1822 428	M. W.	Flax	Yes	1822	PA, Chambersburg	Poss.	* SWS, poss. William J. Major Chester Co., PA
H.A. MACLAHAN	MACLAHAN, H.A.	Winder	Yes				
WILLIAM H. MAIN	MAIN, WILLIAM H.	Patent	Yes	1870	WI, Marietta	Cert.	Pat. # 103906
	MAJOR, WILLIAM J.		No	1829	PA, Downington	Cert.	Chester Co. furn. makers-ad: Spinning wh. Maker
I. MAKEE	MAKEE, I.	Flax	Yes				
S. MARBLE	MARBLE, S.	Wool	Yes		MA, Williamstown	Poss.	poss. "S" is "SIDNEY"
J. MARRILL	MARRILL, J.	Wool	Yes				
	MARROTT, DAVENPORT		No	1800's	PA, Phila.	Cert.	prob. wh. maker in 1800
S. MARSH	MARSH, S.	Wool	Yes		NY	Poss.	barrel tension
W. MARSH	MARSH, W.	Patent	No	1856?			SWS
G. MASON	MASON, G.	Flax	Yes				
CHELTON MATHENY	MATHENY, CHELTON	Patent	Yes	1860	IN, Greensburg	Cert.	#27059(seen)/1860, #62351/1867, #96937/1869
	MATTHEWS, RAWLEY		No	1770's	NC	Prob.	poss. wh. maker in 1770'S
	MAULE, JOSEPH		No	1823	PA, Downington	Cert.	Chester Co. furn. makers-ad for Spinning wh. & reel maker
	MAY, HENRY		No	1806	OH, Chillicothe	Cert.	ads in Scioto Gaz.
	McALISTER, WRIGHT &		No	1770's	NY, NYC	Cert.	poss. wh. makers,see "WRIGHT"
	McCLUIR, JOHN		No	1764	PA	Cert.	run away bond servant ad in Phila. paper
	McCRACKEN, WILLIAM		No	1850	IN, Knox Co.	Cert.	making chairs & wh.

SPINNING WHEEL MAKERS

J. McCULLOUGH	McCULLOUGH, J.				PA	Poss.	
	McDOWELL, NATHANEL		Yes	1763	PA	Cert.	run away bond servant ad in Phila. paper
McINTOSH 1838	McINTOSH,		No	1838			
McINTOSH 1870	McINTOSH,	Flax	Yes	1870	CAN, P.E.I.	Prob.	SWS
ALEXR McINTOSH 18??	McINTOSH, ALEXANDER	Flax	Yes	1809-1820	CAN, Pictou	Prob.	SWS
I.S. McINTOSH 1890	McINTOSH, I.S.		Yes	1890			
McKAY ARKONA	McKAY	Wool	Yes				"ARKONA" stenciled on side also
J. McKAY	McKAY, J.	Wool	Yes		NY	Poss.	Flat posts on tension device
W. McKEAN	McKEAN, W.	Flax	Yes				SWS
	McKEEKEN, DAVID		No	1780's	NC	Poss.	poss. wh. Maker
	McKEEKEN, NATHANIEL		No	1780's	NC	Poss.	poss. wh. Maker
McLEAN	McLEAN	Flax	Yes			Cert.	
	McLEOD, MURDOCK	Patent	No	1874	MI, Grand Haven	Cert.	Pat.# 156230 in 1874
	McNEMAR, RICHARD		No		OH, Union Village	Cert.	Shaker maker but no known examples
D.M.	MEACHAM, DAVID	Wool	Yes		NY, New Lebanon	Cert.	* Shaker Trustee
ERNST MECHELKE	MECHELKE, ERNST	Flax	Yes	1860-1906	WI, Hutisford	Cert.	emigrated from Pomerania
GEORGE MECHIL	MECHIL, GEORGE	Flax	Yes		PA	Poss.	
	MEINECKE, A. & SON		No	1890's	WI, Milwaukee	Cert.	working in 1890
I. MERKEL 1801	MERKEL, JOHN	Flax	Yes	1801	PA, Berks Co.	Prob.	wh. rim 2 long/2 short, olive/shotgun shell spoke, like Humes
J. MERREL	MERREL, J.	Flax	Yes		PA/NJ	Poss.	sax. with upright wh. Supports
	MERRILL, C.	Patent	No	1824,Jul3	NY, Lowville	Cert.	Patent lost, "Spinning machine"(see BATCHELDER & KING)
	MERRILL, GRANVILLE		No	1839-1878	ME, SDL	Cert.	Shaker, poss. Maker
JOSIAH MERRILL	MERRILL, JOSIAH	Wool	Yes	1721-?	NH, Windham	Cert.	
N.M.	MERRILL, NATHAN	Flax	Yes	1806-1809	ME, Alfred	Cert.	Shaker, 1743-1819, Deacon at SDL 1796-1806, Alfred 1806-1809
S.M.	MERRILL, STEPHEN	Wool	Yes	1825-1836	NH, Canterbury	Cert.	Shaker
	MESSENGER, ALFRED	Patent	No	1832,Mar31	MA, Lowell	Cert.	Patent lost, "Spinning wool",(see CALVERT & SOUTHWICK)
	METCALF, GEORGE		No		PA	Poss.	poss. wh. Maker
J. MILES	MILES, J.	Unusual	Yes		CT, Guilford	Poss.	* double flyer type 1, chair wheel
T. MILLARD	MILLARD, THOMAS	Unusual	Yes	1800's	PA, Phila.	Cert.	parlor style
E.J. MILLER 1863	MILLER, E.J.	Flax	Yes	1863	PA/OH	Poss.	
	MILLER, HENRY	Patent	No	1867	MI, Ronald Twp.	Cert.	Pat.# 71897 in 1867
J. MILLER	MILLER, JOEL	Patent	Yes		PA, Springs	Prob.	track wh., flax,wool,winders
	MILLER, SAMUEL		No	1820	IN, Knox Co.	Cert.	working in 1820
JOHN MILNE	MILNE, JOHN	Flax	Yes		CAN.?	Poss.	split lower table,modified "T" top, big maiden turnings, plain spokes
DANIEL MILTIMORE	MILTIMORE, DANIEL	Flax	Yes	1800's	NH, Antrim	Poss.	poss. early 1800'S SWS-"ANTRIM, NH"
MINOR & SESSIONS	MINER, AMOS	Minor's heads	Yes				made heads for $2.00
AMOS MINOR	MINER, AMOS	Minor's heads	Yes	1803/1810	NY, Marcelus	Cert.	* Patent for wool wh.(1803) & head(1810)
GEORGE MITCHELL	MITCHELL, GEORGE	Flax	Yes		NJ	Poss.	
JOHN MITCHELL	MITCHELL, JOHN	Flax	Yes		MA	Poss.	
MIX	MIX	Flax	Yes				
I. MONTGOMERY	MONTGOMERY, JOHN	Unusual	Yes				reg. flax & single flyer on double flyer style
S. MOOR	MOOR, S.	Flax	Yes				

SPINNING WHEEL MAKERS

	Maker		Type	Date	Location	Cert.	Notes
J. MOORE	MOORE, F.D.		Patent	1868	WV, Edray	Cert.	Pat.# 74115 in 1868
	MOORE, J.		Wool			Yes	SWS
M.S. MOREHOUSE & CO.	MOREHEAD, CHARLES A.		Patent	1864	IL, Quincy	Cert.	Pat.# 43327 in 1864
	MOREHOUSE & CO., M.S.		Patent	1867	NY, Cape Vincent	Cert.	made Wight pendulum wh. SWS-"M.S. INITIALS"
	MORGAN, JOHN		Patent	1836,May14	PA, Manayunk	Cert.	Patent lost, "Spinning machine"
S. MORISON	MORISON, S.		Flax, Wool		NY	Yes	sax. has internal crank SWS-"NY"
I. MORRIS	MORRIS, I.		Wool		NY	Poss.	"I" may be "J"
S. MORRISON	MORRISON, S.		Wool			Yes	
?. MOSES	MOSES, ?		Wool			Yes	
	MUELLER, FREDERICK		Wool	1740's	GA, Savannah	Yes	poss. wh. maker in 1740's
	MULLENS, JOHN W.		Patent	1876	KY, Racoon	No	Pat.# 179043 in 1876
I. MUSSELMAN	MUSSELMAN, I.		Flax			Yes	SWS-"I" may be a "J"
I. MYERS	MYERS, I.		Flax		PA, Lancaster Co.	Prob.	
J.N.	N, J.		Flax		PA, W. Cluster	Poss.	
AARON M. NEAL SALEM	NEAL, AARON M.		Flax		(SALEM)	Prob.	
	NELSON, JOHN		Patent	1841,Jan27	OH, Jefferson	Cert.	Patent lost, "Spinning,wool spinner, domestic"
	NELSON, WILLIAM			1754	MA, Boston	Cert.	ads in 1754
	NEWBROUGH, DAVID		Patent	1832,Apr27	OH, RileyTwp.	Cert.	Patent lost, "Spinning wool & cotton"
J. NEWBY 1841	NEWBY, J.		Wool	1841		Yes	
	NEWELL, HUGH			1850	IN, Montgomery Co.	Cert.	wheel maker in 1850
	NEWELL, TIMOTHY			1790's	MA, Sturbridge	Cert.	working 1790'S
F. NEWSWANGER	NEWSWANGER, F.		Flax		PA	Poss.	
	NIXON, FRANCIS M.		Patent	1869	IL, Lena	Cert.	Pat.# 88065 in 1869,table top
	NORCROSS, L.		Patent	1835,Jun15	ME, Dixfield	Cert.	Pat.# 8899 in 1835 table top spindle wh.
C. NORTHRUP	NORTHRUP, C.		Flax				inverted vase leg,table carved & molded,Hepplewhite finial,olive to sg
R. NORTON	NORTON, R.		Flax			Yes	SWS
	NOWELL, PAUL				ME	Poss.	
NUTE'S COMBINED SPINS	NUTE, JOHN HENRY		Patent	1870	CAN.(N.S.)	Cert.	* Can. Patent
L.O. M.D.	O, L.		Flax			Yes	12 spoke,zig zag decoration. see "M.D."
S.I.O. 1768	O, S.I.		Wool	1768	PA	Poss.	
	ODELL, THOMAS G.		Patent	1866	WI, Camp Point	Cert.	also Boyd Glover on Pat.
GEORGE OERTHER	OERTHER, GEORGE		Flax		MI, Monroe	Cert.	last half 19th century
	OLIVER, ANDREW			1796	PA, Chester Co.	Cert.	Chester Co. furn. makers-listed as "wheel maker"
	ORNDORFF, JOHN		Patent	1829,Nov10	KY, Russelville	Cert.	Patent lost, "Spinning wool"
D.L.OSBORNE MAKER	OSBORNE, D.L.		Patent	1868	CAN.?	Yes	"Pat. 1868" like Wight pend.
OSGOOD	OSGOOD					Yes	SWS
	OTIS, RICHARD?				NH, New Canaan	No	
F.O.	OUELLET, FRANCOIS		Saxony	1840-1860	CAN., Quebec	Cert.	Village Des- Aulnaies
J.O.	OUELLET, J.		Saxony		CAN., Quebec	Cert.	brother of "F.O."
AOH 1790	OVERHOLT, ABRAHAM		Flax	1765-1834	PA, Bucks Co.	Cert.	unique quiller dated 1821
A.P.	P, A.		Flax, Wool			Yes	
D.P.	P, D.		Unusual			Yes	double flyer type 1 & flax

SPINNING WHEEL MAKERS

E.P.	P., E.		Wool	Yes			Poss.	horn collar prob. not Shaker
E.P. N.H.	P., E.		Flax	Yes			Poss.	SWS
J.P.	P., J.		Wool	Yes		PA, Haverford Twp.	Poss.	double nut tension, rev. taper table
M.P.	P., M.		Wool	Yes		ME	Poss.	"Sept. 7 1823" on it
M.P.	P., M.		Quiller	Yes	1823			raised letters
M.P.	P., M.		Flax	Yes		ME	Poss.	
M.P.	P., M.		Flax	Yes				
M.V.P.	P., M.V.		Flax	Yes		LA	Poss.	under table tension
R.P.	P., R.		Wool	Yes		MA	Poss.	flower stamps on end
S.P.	P., S.		Flax	Yes				
	PADDLEFORD, PETER		Patent	No	1816,May18	NH, Lyman	Cert.	Patent lost, "Spinning wool & cotton"
F.L. PAED & CO.	PAED, F.L.		Flax	Yes		WI, Green Bay	Cert.	
	PAINE, JOHN		Flax	No	1739-1815	NY, Southhold	Prob.	SWS
PARADIS	PARADIS		Flax	Yes	1875-1940's	CAN. St. Andre, Que.	Cert.	* four generations made wheels
A. PARISH	PARISH, AMBROSE		Flax	Yes	1764-1847	NY, Oyster BayL.I	Cert.	
I. PARISH	PARISH, ISAAC		Flax	Yes	1766-1836	NY, Oyster Bay	Cert.	stamped on top of table
	PARMALEE, ASHEL		Unusual	No		CT, Killingworth	Cert.	Flax, double flyer, "AP" ?
	PARSONS, JOHN		Flax	No	1674-1693	NY, East Hampton	Prob.	SWS
	PATRICK, DANIEL			No		NH, Lyme	Prob.	SWS
	PEABODY, W.H.		Patent	No	1812			Pat.# 1850, no info
	PEARCE, JOHN		Patent	No	1830,Jun30	NY,Yorkshire	Cert.	Patent lost "Spinning machine"
S. PENNOCK	PENNOCK, S.			Yes		PA, Chester Co.	Poss.	
	PENNY, EDWARD		Patent	No	1828,Jun27	NY, Adams	Cert.	Patent lost, "Spinning machine, domestic"
	PERRY, J.S.			No	1865	MN, Forrestville	Cert.	
K. PETERSEN	PETERSEN, KARSTEN		Flax, Wool	Yes	1776-1857	NC, Old Salem	Cert.	
PHELPS & COLLINS	PHELPS		Miner's head	Yes	1845-1928	OH, Grandville?	Poss.	
A. PHILLIP	PHILLIP, A.			Yes				
ALFRED PIERCE	PIERCE, ALFRED		Miner's heads	Yes	1820's	NH, Chesterfield	Cert.	
	PIERCE, ALFRED B.			No		CT, Colchester	Cert.	died 1880
B. PIERCE	PIERCE, BENJAMIN		Wool, heads	Yes	1815-1899	NH, Chesterfield	Cert.	SWS-"heads,wool,related items"
	PIERCE, E.PORTER		Miner's heads	No	1814-1820	NH, Chesterfield	Cert.	
E.P. PIERCE JR.	PIERCE, E.PORTER JR.		Miner's heads	Yes	1785-1865	NH, Chesterfield	Cert.	in Keene, N.H. later
	PIERCE, FRED BENJ.		Miner's heads	No	1845-1928	NH, Chesterfield	Cert.	
	PIERCE, JOHN		Miner's heads	No	1820's	NH, Chesterfield	Cert.	son also working
	PIKE, JEREMIAH			No	1696-1711	MA, Fremingham	Cert.	see PIKE SR.
	PIKE, JEREMIAH JR.			No		MA, Framingham	Cert.	poss. wh. maker (ESP)
	PIKE, JOSEPH		Wool	No	1826	ME, Alfred?	Cert.	son of JEREMIAH PIKE SR.
	PIKE, MOSES			No		MA, Framingham	Cert.	double flyer type 1, flax, wool ,reels
S.P.	PLANT, SOLOMON		Unusual	Yes		CT, Stratford	Cert.	* auto-winding bobbin device on table frame wheel
JOHN PLANTA	PLANTA, JOHN		Unusual	Yes	1798-1802	ENG. Fulneck	Cert.	winder on sax. Base
J. PLATT	PLATT, J.		Winder	Yes				SWS
JACOB PLUM	PLUM, JACOB			Yes				

SPINNING WHEEL MAKERS

			Patent			Cert.	
C. PORTER	POOL, D.			No	PA, Phila.	Cert.	Patent lost "Spinning machine", see COPELAND
	PORTER, ASHEL P			No	NH, Chesterfield	Cert.	Miner's heads ?
	PORTER, C.		Wool,Unusual	Yes	NY	Prob.	* double flyer type 1, barrel tension wool
JAMES PORTER	PORTER, JAMES		Miner's heads	Yes			
N.A. PORTER	PORTER, N.A.		Miner's heads	Yes			
	POTE, J.			Yes	ME	Poss.	Esposito thought Shaker, doubtful
J. POTE							
J. PRAT	PRAT, J.		Winder	Yes			
	PRICE, FRANCIS	1824,Oct5	Patent	No	NY	Cert.	Patent lost,"Spinning wool & cotton"
WILLIAM PRIDE	PRIDE, WILLIAM		Wool	Yes	WV, Kanawch	Poss.	found in log cabin there
	PRINGLE, CASPER	1770's		No	NC	Poss.	working in 1770's
	PURDINGTON, AMOS	1777-1843		No	NH, Weare	Cert.	
	PYLE, JAMES	1748		No	PA, Chester Co.	Cert.	Chester Co. furn. makers-death inventory: lathe,wood for wheels
A.R. 1790	R, A.		Reel	Yes	PA	Prob.	*
D.R.	R, D.		Flax	Yes	NH	Poss.	
H.R.	R, H.		Flax	Yes	MA	Poss.	metal flyer
J.R. 1843	R, J.	1843	Wool	Yes	NY	Poss.	direct drive,sliding table tension,ball top wheel post
J.R.	R, J.	1817	Flax	Yes	IN	Poss.	"1817" left footed treadle
J.R.	R, J.	1807	Wool	Yes		Poss.	"1807" stamped by initials prob. Appla. orgin, unusual curved table
M.R.	R, M.		Unusual	Yes	CAN.	Cert.	double wheel above & below table
P.R.	R, P.		Flax	Yes	CAN.	Poss.	
R.R.	R, R.		Flax	Yes	IN	Poss.	
T.R.	R, T.		Flax	Yes			SWS
HUGH RAMSEY	RAMSEY, HUGH	1754-1831	Flax	Yes	NH, Holderness	Cert.	Several seen
	RAMSEY, JAMES			No	VT, St. Johnbury	Cert.	Esposito
RAMSEY	RAMSEY, WILLIAM	1790's		No	NH, Walpole	Cert.	working 1790'S, Esposito, see CUMMER Pg.134
GEO RANDALL	RANDELL, GEORGE		Flax	Yes			SWS
HARMONIE G.R. 1823	RAPP, GERTRUDE	1823	Flax	Yes	PA, Harmony	Cert.	prob. not a maker but poss. made there
H. RAUCH	RAUCH, H.			Yes	PA	Poss.	
	RAWSON		Wool	Yes	MI, Monroe	Cert.	see ROOT+RAWSON
	READ, DANIEL	1811,Sept10	Patent	No	NY, Brookfield	Cert.	*
LINDSEY READING	READING, LINDSEY		Flax	Yes	PA, Reading	Poss.	questionable name and location
D. REINER	REINER, D.		Flax	Yes	PA	Prob.	seen by Bill Leinbach
J. REINER	REINER, J.		Flax	Yes			SWS
S. REINER	REINER, S.		Flax, Wool	Yes	PA, Leigh...	Prob.	* 6 arm winders- 72" DIA., 150 rev.
	REMINGTON, NATHANIEL	1827,Apr21	Patent	No	NY, Geneva	Cert.	Patent lost "Spinning machine"
REYNOLDS	REYNOLDS		Wool	Yes	ME	Poss.	
J. REYNOLDS	REYNOLDS, J.		Flax	Yes			E-Bay auction
J. REYNOLDS HWP 1800	REYNOLDS, J.	1800?	Flax	Yes	MA, Cape Cod	Poss.	many seen,SWS-Boston
JOHN REYNOLDS	REYNOLDS, JOHN		Wool	Yes			N"s are backwards
R.J. REYNOLDS	REYNOLDS, R.J.		Flax	Yes	MA	Poss.	
	RHODES, FREDERICK	1770's		No	NC	Prob.	poss. maker working in 1770'S
	RICE, JAMES	1869	Patent	No	IN, Prairie Creek	Cert.	Pat.# 96619

SPINNING WHEEL MAKERS

	Name	Type	Patent	Date	Location	Cert.	Notes
	RICHARDSON, JAC.	Patent	No	1816, Oct16	NY, Scipio	Cert.	Patent lost, "Spinning wool & cotton"
SR AL	RING, SAMUEL	Flax, Wool	Yes	1766-1848	ME, Alfred	Cert.	* Shaker Trustee from 1809-1814
	ROBINSON, GEORGE W.	Patent	No	1826, Mar16	RI, Providence	Cert.	Patent lost, "Spinning wool &cotton by hand"
J. ROGERS	ROGERS, J.	Flax	Yes			Cert.	
	ROGERS, WM. P.	Patent	No	1826, Jun15	OH, Harpersfield	Cert.	Patent lost, "Spinning wool", see CORRILL, J.
ROOT+RAWSON MONROE	ROOT	Wool	Yes		MI, Monroe	Cert.	
B. ROOT	ROOT, B.	Winder	Yes				B" could "E
L.B. ROOT	ROOT, L.B.	Wool	Yes		MI, Monroe	Cert.	has ROOT+RAWSON label "L" poss. "E"
S. ROSE	ROSE, S.	Flax	Yes		CT, Stratford	Poss.	* also reel
	ROUNDTREE, ANDREW	Wool?	No	1700's	TN	Cert.	late 1700's, wheels, 7 looms
	ROWE, JAMES		No				SWS
	ROWE, JONAS H.	Patent	No	1867	NY, Hudson	Cert.	Pat.# 70622 in 1867, lever action
	ROWELL, JOHN		No		NH, Andover	Cert.	poss. Maker
J. ROY	ROY, J	Winder	Yes				4 armed, unusual style, nicely done, stamped both ends
J.Z. RUST	RUST, R.Z.	Miner's head	Yes				SWS
P. RUTH	RUTH, P.	Flax	Yes				
A.S.	S, A.	Unusual	Yes				double flyer type 1 poss. STURTEVANT?
A.S. 1835	S, A.	Wool	Yes	1835	NH	Poss.	Single nut barrel tension
D.S. 1823	S, D.	Flax	Yes	1823, 1830	PA, Berks Co.	Poss.	SWS, 1830 seen also
D.S.	S, D.	Unusual	Yes		CT	Prob.	American castle
D.B.S.	S, D.B.	Flax	Yes				
E.S.	S, E.	Wool	Yes		NY	Poss.	barrel tension
F.S.	S, F.	Flax, Wool	Yes		NY	Poss.	barrel tension
G.S.	S, G.	Flax	Yes				
H.S.	S, H.	Flax	Yes				"New Eng. style"
H.S. 1863 2	S, H.	Flax	Yes	1863, 1865			SWS, 1865 seen also
I.S.	S, I.	Unusual	Yes				double flyer type 1
I.S.	S, I.	Swift	Yes		PA	Poss.	squirrel cage style
I.S. 415	S, I.		Yes				poss. deep south
I.C.S.	S, I.C.	Flax	Yes				"I" could be "J"
	S, J.	Unusual	Yes		NY, Rochester	Poss.	Farnham type
J.S.	S, J.	Flax	Yes		PA	Poss.	
J.S.	S, J.	Flax, Wool	Yes		VT	Poss.	
J.S. 1843	S, J.	Flax	Yes				date picked in, "JS" stamped in
	S, M.	Wool	Yes		MA	Poss.	two nut rod tension above table
	S, N.	Wool	Yes		NH	Prob.	Shaker
O.I.S.	S, O.I.	Flax	Yes				SWS
P.S.	S, P.	Flax	Yes		PA	Prob.	
S.S.	S, S.	Flax	Yes				plain
S.S. G.	S, S. G.	Flax	Yes		NY	Poss.	on end & top
S.S.	S, S.	Flax	Yes		NY	Poss.	Hepplewhite finial
T C.S.	S, T C.	Flax	Yes		NY	Poss.	

SPINNING WHEEL MAKERS

		Wool/Flax		Date	Location	Cert.	Notes
T.L.S.	S, T.L.	Wool	Yes		MA		SWS
?.C.S.	S,?,C.	Flax	Yes			Poss.	1st letter poss.. "A","C",or"T"
SALDRICH	SALDRICH	Wool	Yes				prob. S. ALDRICH
B. SANFORD	SANFORD, BEARDSLEY	Unusual	Yes	1790-1868	NY, Fergusonville	Cert.	* flax, wool, Webster type, double flyer type 1
D. SANFORD	SANFORD, DANIEL?	Wool	Yes	1734-1815	NY, Southampton	Prob.	prob. broken "B" in a B. Sanford stamp but Daniel was his father
E.B. SANFORD	SANFORD, ELIAS BRISTOL	Flax,Unusual	Yes	1791-1881	CT, Newtown	Prob.	* 1816 patent, double flyer type 3, saxony, (father is Issac)
HENRY SANFORD	SANFORD, HENRY	Unusual	Yes				double flyer type 1
I. SANFORD	SANFORD, ISAAC	Unusual	Yes		CT, Newtown	Prob.	* double flyer type 1 (son is E.B. Sanford)
J. SANFORD	SANFORD, J.	Wool,Unusual	Yes		CT	Poss.	* double flyer type 1, is "J" I"?
S.S.	SANFORD, SAMUEL?	Unusual	Yes		CT	Prob.	Leadbeater book photo
W. SAREINS	SAREINS, W.	Flax	Yes				
	SAUL, JOSEPH		No	1754	PA	Cert.	run away bond servant ad in Phila. paper
	SAUL, JOSEPH	Flax	No	1750	PA, Phila.	Prob.	SWS
	SAWYER, ELBERT	Patent	No	1877	KY, Pond Fork	Cert.	Patent #193562 with John Smith
D.S.	SAYRE, DAVID	Flax	Yes	1773-1819	NY, Bridgehampton	Prob.	
	SCHARFER, P.	Flax	No		MN, St. Peter	Prob.	SWS
	SCHARFER, P.		No		MN, St. Peter	Cert.	
BY SCHNOOVER	SCHNOOVER,	Flax	Yes				SWS
B. SCHREIBER	SCHREIBER, B.	Wool	Yes				
J. SCHUMM	SCHUMM, J.	Flax	Yes		PA, Lebanon Co..	Poss.	seen by Bill Leinbach
Wm. SCOTT	SCOTT, Wm.	Flax	Yes		PA/OH	Poss.	picked in
H. SEIP	SEIP, H.	Flax	Yes				
I. SELLERS	SELLERS, I.	Flax	Yes		PA, Bucks Co.	Poss.	
P. SELLERS	SELLERS, P.		Yes				
H.H. SEYMOUR	SEYMOUR, H.H.		Yes				
	SHARP, JOHN	Patent	No	1822,Mar13	NY, Whitestown	Cert.	Patent lost "Spinning wool & yarn"
	SHAW, AARON		No	1806	PA, Plumstead	Cert.	ad in 1806
D. SHAW	SHAW, D.	Flax	Yes				
JACOB SHAW JR.	SHAW, JACOB JR.	Patent	Yes	1848	OH, Hinkley	Cert.	* Pat.# 5847 in 1848, lever action
	SHEARMAN, JOHN		No	1795	PA, Phila.	Cert.	AWC- wheel maker, William Cox was his apprentice
J. SHELL	SHELL, J.	Wool	Yes				
D. SHELLY	SHELLY, D.	Flax	Yes		PA	Prob.	several seen
	SHERMAN, JOHN		No				poss. wh. maker
	SHOEMAKER, JACOB		No	1738	PA	Cert.	run away bond servant ad in Phila. paper
L. SKEELS	SKEELS, L.	Flax	Yes				
E. SKINNER	SKINNER, BARTON	Flax, Wool	No	1860's	NH, Chesterfield	Cert.	made wh. & heads during Civ. War
	SKINNER, ELIJAH	Patent	Yes	1786-1871	NH, Sandwich	Cert.	Pat.# 2895 1/2 in 1818
I. SLATER	SLATER, I.		Yes		PA, Berks Co.		
	SLATER, SAMUEL	Patent	No	1825,Apr4	VT, Miiddletown	Cert.	Patent lost "Spinning machine"
?SLEEPR	SLEEPR	Wool	Yes				bone collar, backwards legs
A. SLOAN	SLOAN, A.	Flax	Yes		MA	Poss.	
J. SMAIL	SMAIL, J.	Flax	Yes				E-Bay auction

SPINNING WHEEL MAKERS

E. SMITH	SMITH, E.		Flax		VT	Poss.	
I. SMITH	SMITH, I.		Flax	Yes	CT	Poss.	
	SMITH, JOHN & EGBERT		Patent	No	1877	KY, Pond Fork	Cert.	Pat.# 193562
	SNAPP, JULIA E.		Patent	No	1867	IL, Georgetown	Cert.	Pat.# 180387, attached to sewing machine
D. SNELL	SNELL, D.		Flax	Yes		PA	Cert.	seen by Bill Leinbach
M. SNYDER	SNYDER, M.		Wool	Yes		PA	Prob.	massive, sliding table, red & green stripes
	SOLL, JOSEPH			No	1750's	PA, Phila.	Prob.	poss. wh. maker in 1750'S, Esposito
D. SON DOVER SCH	SON, D.		Flax	Yes				
	SOUTHWICK, R.		Patent	No	1832,Mar31	MA, Lowell	Cert.	Patent lost,"Spinning wool",(see Calvert & MessengerR)
	SPEARS, CHETNEY		Flax	No	1820's	IN, Brownsville	Cert.	quillers,winders,See "CHURCH"
E. SPENCER	SPENCER, E.		Flax	Yes		CT	Poss.	
	SPINNEY, JOHN		Flax, Wool	No	1762?	NH?	Poss.	SWS
P + SPOIR	SPOIR, P		Wool	No	1850-1880			SWS
	SPRINKEL, JAC.		Patent	No	1816,Aug23	VA, Wythe	Cert.	Patent lost,"Spinning wool & cotton"
B. SQUIER WARRENTED	SQUIER, B.		Flax	Yes				"B" poss.."R"
I. STABB	STABB, JOHN		Flax, Wool	Yes	1820's	PA, Williamsport	Cert.	
E.W. STANGER	STANGER, E.W.		Flax	Yes				
J.W. STANTON	STANTON, J.W.		Wool	Yes				single post barrel tension device
J. STAUB	STAUB, J.		Flax	Yes		PA	Prob.	
JAB STAUB	STAUB, JAB		Flax	Yes		PA/OH	Prob.	"B" is raised
J. STEALY	STEALY, J.		Flax	Yes		PA/MD	Prob.	
J. STEBELTON	STEBELTON, J.		Flax	Yes				
	STEWART, HUGH			No	1824	PA, Greensburg	Cert.	ads in 1824 for wh. & reels
R. STEWART	STEWART, R.					MA	Poss.	
W.M.STEWART	STEWART, W.M.		Wool, Unusual	Yes				SWS, Irish castle
J. STOCKING	STOCKING, J.		Wool	Yes		NY	Prob.	"N" reversed, barrel tension
T. STOCKING	STOCKING, T.		Winder	Yes				
	STODDARD, W.			No	1750's	NY, Oyster Bay, L.I.	Prob.	
JOHN STOHER	STOHER, JOHN							
E. STONE	STONE, E.		Flax	Yes				
E. STONE WARRENTED	STONE, EBENEZER		Wool	Yes		NY	Poss.	barrel tension
F. STONE	STONE, F.		Flax	Yes				
N. STONE	STONE, N.		Flax	Yes				
S. STONE	STONE, SILAS ?		Flax	Yes	1821	NY, Hudson	Cert.	poss. SILAS
	STONECIPHER, DANIEL		Wool	Yes		TN, Morgan Co.	Cert.	wheel and loom
	STRAIN, JOSEPH		Patent	No	1871	CAN, Artemesia	Cert.	Pat.# 113594
	STROBRIDGE, GEORGE		Wool	No	1820	NY	Poss.	SWS-Esposito
JOHN STURDEVANT	STURDEVANT, JOHN		Unusual	Yes	1830's	NY, Albany/Utica	Prob.	"CRACON" on it also,flax, N.Y. style wool
E.S.	STURTEVANT, ELI (?)		Flax	Yes		CT, Bridgewater	Prob.	
	SUGAR, DAVID JR.			No	1738-1820	PA, Chester Co.	Cert.	Chester Co. furn. makers- sm. wheel in inventory in 1774
I. SUGAR	SUGAR, I.		Flax	Yes		PA/OH	Poss.	Several seen
W. SEELY	SWEELY, W.		Flax	Yes				SWS

SPINNING WHEEL MAKERS

		Patent	No			Cert.	
M. SYNDER	SYKES, WM.	Wool	No	1834,Mar10	MD, Fredricktown		Patent lost,"Spinning wool",(see CONRADT)
T	SYNDER, M.	Flax	Yes		PA	Poss.	line over "T"
A.T.	T, A.	Flax, Unusual	Yes		MA	Poss.	* double flyer type 2 (T-top) flax is prob. the same maker)
A.T. 1808	T, A.	Flax	Yes	1808	PA	Poss.	dated"1808"
D.T.	T, D.	Unusual	Yes		CT/MA	Poss.	double flyer type 2 (T-top)
I.T.	T, I.	Flax	Yes	1849	PA/OH	Poss.	Several seen
J.T.	T, J.	Wool	Yes				stamp raised
J.T.	T, J.	Flax	Yes				
J.T.	T, J.	Flax	Yes	1815			heart design-not makers
M.T.	T, M.						SWS
R.T. NO.44 RT 1829	T, R.	Flax	Yes	1829	NY	Prob.	S. Appl. highlands
S.T.	T, S.	Wool	Yes			Prob.	Barrel tension
S.S.T.	T, S.S.	Flax, Wool	Yes	1828+1835	PA	Poss.	
W.O.T. 1828	T, W.O.	Flax	Yes		PA	Prob.	picked in
S. TAYLOR	TAYLOR, S.	Flax, Wool	Yes		NY	Prob.	barrel tension
TEASDALE	TEASDALE, ?	Unusual	Yes	1600's	ENG, Derbyshire	Cert.	* Girdle wheel, paper label
DAVID TETER	TETER, DAVID	Patent	Yes	1865	IA, Batavia Sta.	Cert.	Pat. # 47685 in 1865, other names on wh.
THAYER	THAYER	Flax	Yes				E-Bay auction
T. THAYRE	THAYRE, T.	Wool	Yes		MA, Deerfield	Prob.	wood axle, holed tension,chamfered legs, SWS-"DEERFIELD"
C. THOMAS 1832	THOMAS, C.	Flax	Yes	1832	PA	Poss.	
D. THOMAS 1824	THOMAS, D.	Flax	Yes	1824	PA, Bucks Co.	Prob.	reel & several flax seen
D.H.THOMAS	THOMAS, D.H.	Flax	Yes		IL	Poss.	
L. THOMAS 1832	THOMAS, L.	Flax	Yes	1832	PA, Lancaster	Poss.	
T. THOMPSON	THOMPSON, T.	Flax	Yes		MA	Poss.	several seen
H. THOMSON	THOMSON, H.	Flax, Wool	Yes		ME/NH	Prob.	* quiller too, not Shaker
J. THOMSON	THOMSON, J.	Flax	Yes				
	THORPE, JOHN	Patent	No	1828,Nov25	RI, Providence	Cert.	Patents lost, "Spinning machine"also 12/31/1828,1/23/1829,6/13/1829
MOSES TIC?ENOR	TIC?ENOR, MOSES	Wool	Yes			Poss.	Several seen
TILDEN	TILDEN,	Flax, Wool	No	1780-1800	PA, Haverford Twp.		chairmaker who made wh.
A. TODD	TODD, A.	Flax	Yes		PA, Phila.	Prob.	*
R. TODD	TODD, R.	Flax	Yes		PA, Bucks Co.	Prob.	SWS-"R" could be a "B"
W. TONG	TONG, W.	Flax	Yes				
	TOTTON,LAFFORD	Patent	No	1824,Nov13	NY, Schoharie	Cert.	Patent lost,"Spinning wool, domestic", 11/22/1826 also
	TOWN, ELISHA	Patent	No	1812,May6	VT	Cert.	Patent lost "Spinning machine" (see BALDWIN)
R. TOWNLEY 3 DRYDEN	TOWNLEY, R.	Wool	Yes	1838	NY, Dryden	Prob.	1838 82 25 2" on spindle post
I. TRACY	TRACY, I.	Flax	Yes		OH	Poss.	marked on side
	TRAUGER, CHRISTIAN		No		PA, Bucks Co.	Prob.	SWS
J. TREMBLE	TREMBLE, J.	Flax	Yes		MA		
TROND	TROND	Wool	Yes			Poss.	horn collar
	TROWBRIDGE, GEORGE		No	1820's	NY, Solon	Cert.	Esposito-working in 1820'S

SPINNING WHEEL MAKERS

Label/Stamp	Maker	Type	?	Date	Location	Certainty	Notes
JOHN TROXEL 1857	TROXEL, JOHN	Unusual	Yes	1857	OH	Prob.	double wheel, double treadle
J. TRUESDEL	TRUESDEL, JESSE	Flax, Wool	Yes	1820-1830	NY, Newark Valley	Prob.	branded on top, owned Joel Farnam's shop
D. TUTTLE	TUTTLE, D.	Wool	Yes				
JONATHON TYSON	TYSON, JONATHON	Flax	Yes		PA, Phila.	Prob.	label & stamp
S.U.	U, S.	Flax	Yes		PA	Poss.	"S" with. Dutch style, thick rim, ball turnings
A.U.	U, A.	Wool	Yes				
	UREY, JOHN		No	1850	IN, Carroll Co.	Cert.	working in 1850
	VERNON, JAMES		No	1770's	NC	Poss.	working in 1770'S
	VIGEANT, SERAPHIN	Flax	No	1880	MA, Holyoke	Prob.	SWS-"Canadian Patent holder"
	VINCENT, JOHN	Heads	Yes		OH, Vincent	Cert.	Made Minor's heads
(NOT MARKED)	VINCENT, JOHN CALEB	Wool	Yes	1863	OH, Vincent	Cert.	barrel tension, heads, winders
	VOEGTLI, FRANZ	Patent	No	1867	MO, Montgomery	Cert.	Pat.# 69728/1867, 111991/1871, 147200/1874, 246251/1881
R.J.VOLLAN	VOLLAN, R.J.	Flax	Yes				
A.W.	W, A.	Flax	Yes				stamp on distaff
A.W.	W, A.	Wool	Yes		CT	Prob.	Shaker like but narrow rim
C.W.	W, C.	Flax	Yes				
C.W.	W, C.	Wool	Yes		MD/VA	Poss.	sliding table, mark on top
C.W. 1823	W, C.	Flax	Yes				SWS
D.W.	W, C.	Flax	Yes		PA	Prob.	stamped on top
D.W. 1844	W, D.	Flax	Yes	1844			prob. picked in, some paint
E.W.	W, E.	Unusual	Yes		PA	Prob.	4-leg upright flax, see "E.M."
E.S.W.	W, E.S.	Flax	Yes				3/4" letters stamped on top of table
F.W. 1814	W, F.	Flax	Yes	1814			internal crank, wh. supports connected AT top
G.W.	W, G.	Unusual	Yes		CT, Guilford	Prob.	* Chair wheel, square legs, spoked wheels
H.W.	W, H.	Wool	Yes				
I.W.	W, I.	Winder	Yes		ME	Poss.	6-arm winder
I.W.	W, I.	Flax	Yes			Poss.	"T" stretcher
J.W.	W, J.	Flax	Yes		PA	Poss.	
J.B.W.	W, J.B.	Unusual	Yes				double flyer type 1
	W, P.	Flax	Yes		PA	Poss.	
WW-623-41	W, W.	Flax	Yes	1841?	PA	Poss.	"W"'s overlap
W.W.	W, W.	Flax	Yes		MA	Prob.	*
H.R. WAGGONER	WAGGONER, H.R.	Flax, Wool	Yes	1800-1850	PA, Lancaster	Prob.	sliding table on wool, SWS-dates
	WAIT, JUSTIN	Patent	No	1862-1871	WI, Watertown	Prob.	SWS
WALTER B. WALKER	WALKER, WALTER B.	Patent	Yes	1874	KS, Cherokee Sta.	Cert.	Pat.# 154969 in 1874, accordian style
	WARBURTON, JOHN		No	1790's	CT, E. Hartford	Prob.	working in 1790'S
F. WARD	WARD, F.	Flax	Yes				
S. WARD	WARD, S.	Winder	Yes				
S. WARNER	WARNER, S.	Wool	Yes		NY	Poss.	raised stamp
E. WASHBURN	WASHBURN, E.	Flax	Yes		ME	Poss.	
T. WATERS	WATERS, T.	Winder	Yes				
H. WATT	WATT, H.	Flax	Yes				

SPINNING WHEEL MAKERS

J. WEAVER	WEAVER, J.	Unusual	Yes		NY	Poss.	Farnham style
A. WEBSTER	WEBSTER, ALPHEUS	Patent	Yes	1810,Mar31	NY, Green Co.	Cert.	* "Spinning mach..",1812,DEC21 "Spinning flax & hemp", other style
S. WEEKES	WEEKES, S.	Flax	Yes		PA.	Poss.	from William Ralph
	WEEKS, CAPT.		No	1722	MA, Goshen	Cert.	
	WELLS, MICAH	Flax	No		NY, Aquebogue	Prob.	SWS
J. WERDEN	WERDEN, J.		Yes				
E. WETTZEL	WETTZEL, E.	Flax	Yes		PA	Cert.	like Phila. Saxony light stamp,poss.."J. WEITZEL"
A. WHEELER	WHEELER, A.	Wool	Yes		ME/NH	Poss.	one stamped, one stenciled
HIRAM F. WHEELER	WHEELER, HIRAM F.	Patent	Yes	1838/1846	PA, Springfield	Cert.	Pat.# 710 & 4892 in '38 & '46(seen)
	WHEELER, J.R.&J.B.	Patent	No	1825,May26	NY, Kingsbury	Cert.	Patent lost "Spinning machine", also 1826,Nov.3, & J.B. in 12/30/29
	WHEELER, ZACCHEUS	Patent	No	1816,Jul18	NH, Richmond	Cert.	Patent lost "Spinning machine"
	WHITE, SIDNEY	Patent	No	1824,Dec24	MA, Wrentham	Cert.	Patent lost "Spinning machine"
WHITELAW	WHITELAW	Wool	Yes		ME	Poss.	Shaker like
LOVE & WHITELOCK	WHITELOCK, FRANKFORD	Flax, Wool	Yes	1790's	PA, Phila.	Cert.	NY style wool
E. WHITMAN	WHITMAN, E.	Flax	Yes				"N" is backwards, stamped twice
WHITNEY	WHITNEY				PA	Poss.	
W. WICKES	WICKES, W.	Flax	Yes				stamped in
WIGHT'S PENDULUM	WIGHT, LYMAN	Patent	Yes		PA, Benton	Cert.	* Pat.# 14482, 2 types ,WS. & NY
ASHER-WILCOX	WILCOX, ASHER-	Unusual	Yes				Double flyer type 1, see "ASHER"
WILDER	WILDER	Miner's head	Yes				prob. NH WILDERS
	WILDER, ABIJAH		No	1752-1835	NH, Keene	Prob.	SWS
A+A WILDER KEENE NH	WILDER, ABIJAH JR.	Wool	Yes	1784-1864	NH, Keene	Cert.	SWS-ABIJAH & dates
AZEL WILDER KEENE NH	WILDER, AZEL	Flax, Wool	Yes	1788-1860	NH, Keene	Cert.	SWS-flax & dates
J. WILEY 1560	WILEY, J.	Flax	Yes				is "1560" a date or?
	WILLIAMS		No		PA, Canton		
E.S. WILLIAMS	WILLIAMS, ENOCH SLOSSEN	Unusual	Yes	1781-1855	NY,Tioga	Cert.	* trained with Farnham, made similar wh.
W. WILLIAMS	WILLIAMS, W.	Flax	Yes				SWS
WILLSON 1824	WILLSON	Flax	Yes	1824			
D. WILSON 1812 NO. 306	WILSON, D.	Flax	Yes	1812	PA/OH	Prob.	
WILSON &F	WILSON, DICKSON	Patent	Yes	1866	IL, Adams Co.	Cert.	* Pat.# 58015 in 1866,HENRY FAIRBANKS is "F"
	WILSON, HENRY	Patent	No	1818,Dec29	NY, Mendon	Cert.	Pat.# 3054(1818) also "Spinning Machine"1827Jul 13,Pomfret,NY
I. WILSON	WILSON, I.	Flax	Yes		PA/MD	Poss.	T- stretcher
W. WILSON 1824	WILSON, W.	Flax	Yes	1824	PA	Poss.	
	WILTSE, A.S.	Patent	No	1825,Jun28	NY, Tioga	Cert.	Patent lost, "Spinning wool & cotton"(see Farnham)
I. WIN	WIN, I.	Wool	Yes				winder also
J. WINE	WINE, J.						
F.W.	WINKLEY, FRANCIS	Flax, Wool	Yes	1807-1828	NH, Canterbury	Cert.	* Shaker Trustee 1807-1828
WISHENDORF & SU???	WISHENDORF,		No	1867	WI, Fond Du Lac	Prob.	SWS
	WITHERED, J.	Patent	No	1836,Mar30	MD, Baltimore	Cert.	Patent lost, "Spinning wool"
WITTSE AND FARNHAM	WITTSE, A.S.		Yes	1825	NY, Owego	Prob.	SWS
	WOERHEIDE, JOBST	Flax	No	1840-1850	MO, St. Louis	Prob.	SWS
WOOD	WOOD,	Flax	Yes				SWS

SPINNING WHEEL MAKERS

			Tool					
C. WOOD	WOOD, C.		Patent	Yes		NY, Lowville		SWS
	WOOD, ELIAS			No	1824, Sept3		Cert.	Patent lost, "Spinning machine, domestic"
G. WOOD	WOOD, G.		Flax, Wool	Yes				stamped on top & side
M. WOOD	WOOD, M.		Unusual	Yes		NH	Poss.	AM. style Irish castle
P.WOOD	WOOD, PHINEAS		Flax, Wool	Yes	1766-1846	VT, Wyndom Co.	Poss.	barrel tension
WOODFORD & HEIST	WOODFORD,			Yes	1869	WI, Madison	Prob.	SWS
JOHN C. WOODRUFF	WOODRUFF, JOHN C.		Flax	Yes				raised stamp
J. WOODWARD	WOODWARD, J		Flax	Yes				12 spokes
	WOOSTER, D.S.		Patent	No	1831			Pat.# 6380, no info
	WOOSTER, DAVID S.		Patent	No	1831, Feb14	NY, Sheldon	Cert.	Patent lost, "Spinning wool"
	WRIGHT & McALISTER			No	1770's	NY,NYC	Cert.	poss. wh. makers , see "McALISTER"
Z.W.	WRIGHT, ZADOCK		Flax, Wool	Yes	1793-1807	NH, Canterbury	Cert.	* Shaker Trustee 1793-1807
JONA WYMAN	WYMAN, JONA		Flax	Yes				SWS
M.I.X.	X, M.I.		Flax	Yes				see "MIX"
I.Y.	Y, I.		Flax	Yes		NY, L.I.	Poss.	similar to I.PARISH, "12" on one
I.Y.	Y, I.		Flax,Wool	Yes				black letters
J.Y.	Y, J.		Flax	Yes		NY, L.I.	Poss.	similar to I.Y.
M.J.Y. No. 383	Y, M.J.		Unusual	Yes		PA	Prob.	* nos. seen: 278, 381, 383, 385
T.Y.	Y, T.		Wool	Yes		NY,L.I.	Poss.	2 nut rod tension, stamp like "IY"
D. YODER	YODER, D.		Tow wheel	Yes				
F. YOUNG	YOUNG, F.		Flax	Yes		CAN.?	Poss.	
J. YOUNG	YOUNG, J.		Flax	Yes		CAN, Lunenberg	Cert.	
G. YUNKER	YUNKER, G.		Flax	Yes		PA/OH	Prob.	
	ZUBLIN, DAVID			No	1803 Died	PA, Chester Co.	Cert.	Chester Co. furn. makers- made wool wheels
	ZUBLIN, JOHN			No	1847 Died	PA, Chester Co.	Cert.	Chester Co. furn. makers- listed as a turner

Glossary

accelerating head: removable spindle assembly for a wool wheel, including spindle and an accelerating wheel. Spindle commonly held on by braided corn husks.

accelerating wheel: small intermediate wheel with whorls (pulleys) used to increase the efficiency of the drive wheel.

bat's head: a removable spindle assembly for a wool wheel, which includes a spindle and a wooden piece shaped like a paddle.

bobbin: spool. Bobbins on spinning wheels have a whorl (pulley) which the drive cord sets in motion. Bobbins used by weavers for warping their looms are usually larger and have no whorls.

bobbin winder: used to wind thread onto a bobbin (or a quill) for use by a weaver; Many bobbin winders are about half the size of wool wheels and often resemble them, but they usually have a small handle either or on a spoke or as a crank on the axle.

boudoir wheel: found in Germany and England these wheels are small wheels on tall ungainly, often detachable, legs. They often have a small drawer under the table and may have a hand crank in front as well as a treadled crank in back. They are sometimes painted with vines and flowers. The small drive wheels are metal or have a metal insert to give necessary momentum for treadling.

brake: a retarding band used to slow down a bobbin or a flyer to get needed speed differential so that thread will wind onto the bobbin as it is being spun; also short for flax brake which was used in the processing of flax prior to its being spun.

cards: small wire teeth set in leather and fastened to wooden paddles, used to straighten wool or cotton fibers before spinning in a domestic setting. Machine cards were later incorporated into large machinery run by waterpower at a carding mill.

carriage wheel: see boudoir wheel.

castle wheel: a flax wheel with the drive wheel directly above the bobbin and flyer unit. Found in Ireland and Scotland, they are best known in the Irish version, which heavily influenced makers in the Lancaster, Pennsylvania area. When we first started collecting, some people used castle wheel to refer to any upright flax wheel.

click reel: reel which makes a "snap" or "click" after a fixed number of revolutions.

clock reel: reel which not only clicks but also has a dial and a rotating pointer which indicates how much thread has been wound on.

combs: see wool combs.

clothesline wheel: see "Pleasant Spinner."

direct drive: spindle holder with maidens and mother-of-all for a wool wheel. It may be removable, but most are not.

distaff: a stick for holding the long flax fibers while spinning. Distaffs could be attached to the spinning wheel, free-standing, or stuck into the belt or girdle of the spinner.

drafting: the spinner pulls the fibers apart as spinning occurs such that an even thread is formed as the wheel puts the twist on the fiber.

drive wheel: the wheel on a spinning wheel or winder which gives power to the tool.

flax brake: implement consisting of parallel strips of wood on a pivot which close together and help break up the flax such that flax fibers are freed for spinning.

flax wheel: spinning wheel with a bobbin and flyer mechanism. It typically has a treadle. By the late 19th century they were adapted to spin wool by making the flyer orifice and hooks larger.

flyer: U-shaped arms attached to a metal spindle with a whorl. The arms are normally of wood. The flyer whorl is detachable and a removable bobbin, also with a whorl, fits between the closed end of the U and the flyer whorl. The flyer has "teeth" (hooks) which are used to guide the spun thread onto the bobbin in an orderly fashion. The end of the flyer facing the spinner has an opening, the orifice.

hackle: a board full of nails or sharp wooden spikes arranged in rows and columns for use in flax processing, often they have a box to cover the nails. Also spelled "hetchel."

horizontal wheel: flax wheel in which the drive wheel is horizontal to the bobbin and flyer.

lazy kate: holder for bobbins which may be used to facilitate plying.

leathers: leather pieces which hold the bobbin and flyer assembly or spindle.

linen wheel: old term for flax wheel.

maidens: the two vertical posts between which the spindle or bobbin and flyer rests.

maker's mark: names or initials stamped, burned, or picked with a sharp tool into the wheel to identify the maker.

Minor's head: detachable accelerating wheel-head used on wool wheels to increase spindle efficiency. "Minor" refers to the inventor Amos Miner, whose name is commonly found misspelled on labels attached to the head.

mother-of-all: turned cross piece which holds "maidens."

niddy noddy: a hand reel.

parlor wheel: a small upright flax wheel with a table and the bobbin and flyer above the drive wheel.

pencil roving: machine carded wool in a continuous coil, which is the diameter of a little finger.

Picardy wheel: a saxony-like flax wheel from the Picardy area in France with an unusual bobbin and flyer system in which the bobbin and flyer are both in front of the maidens. Sometimes these wheels have only a hand crank and no treadle. Not surprisingly, versions of this wheel made their way to French Canada. In the late 19th century this type of bobbin and flyer unit was used on parlor wheels from central Europe.

"Pleasant Spinner": The name given by Daniel Read to his patent wheel. Spindle wheels which have no drive wheel. Power is supplied to an accelerating wheel by means of a long loop of cord. Many types exist.

quill winder: see bobbin winder.

reel: textile tool used to wind off the thread from the spinning wheel. They are of a standardized size, so that they may be used for measuring the length of thread.

reeling pin: small, turned wooden pin used to guide linen thread onto reel. The pin sat in a hole in either the wheel or the reel when not in use. Usually only found in wheels from eastern Pennsylvania.

ripple: looking like a metal comb for a giant, it had tangs on each end so that it could be driven into a stump or block of wood. Flax was pulled through it and the seeds removed for future planting or further processing into linseed oil. Then the flax could be further processed with brakes and hackles and made suitable for spinning into linen thread.

rolag: bundle of wool or cotton fibers prepared for spinning by hand cards.

roving: fibers prepared for spinning by machine into a continuous coil.

saxony: the style of flax wheel most commonly seen in the United States. Slanted table, treadle wheel with bobbin and flyer.

slapper: the flexible arm on a reel which slaps against the side and indicates a fixed number of rotations of the reel's arms.

spinning jenny: multiple spindle machines that kicked off the Industrial Revolution, a few of these were designed with 10-12 spindles for home wool spinning use. Improperly used by Wallace Nutting to designated a Connecticut chair wheel, old time dealers used to call flax wheels spinning jennies. "Jenny" is a corruption of the word *engine*.

spool knave: see lazy kate.

stock: see table.

swift: swifts are used to hold thread when it is in a skein so that it can be more easily handled by the textile worker. There are several different types. The three most common are the basket swift, the squirrel cage swift, and the umbrella swift. Swifts are really "unwinders."

table: the thick board which provides platform for a spinning wheel. It is the receiver for the legs and the wheel supports. Also called the stock or bench by some writers.

triller: a swift.

upright wheel: any flax wheel in which the drive wheel is above or below the bobbin and flyer.

vertical wheel: upright wheel.

wheel-less wheel: see "Pleasant Spinner."

wheel "boy" or "finger": small hand-held wooden piece with knob on the end used to turn the drive wheel on a wool wheel.

wheel post: the support or supports from which the drive wheel is suspended.

whorl: the pulley on the end of a bobbin or attached to the flyer shaft. Sometimes spelled "whirl."

weasel: click reel. Supposedly, mothers turned the click reel reciting the nursery rhyme, which ends "pop goes the weasel," and hence the association. The story is likely apocryphal.

winder: some use winder as a synonym for reel and others only for bobbin or quill winders.

wool combs: two or more rows of long needle-like nails set in paddles, used by professional woolen combers to prepare the very long stapled wools for worsted spinning.

wool wheel: a spinning wheel with a spindle. It was commonly used to spin wool or cotton. The drive wheels are often quite large (42-44 inches). The wheels were turned by hand.

yarn winder: see reel.

yarn: spun wool; often used synonymously with thread.

yarn count system: a system of measurements indicating length and diameter of spun thread so that weavers and spinners can communicate.

Bibliography

Andrews, Edward D. *The Community Industries of the Shakers: New York State Handbook Number 15*. The State University of New York, 1933.

Asplundh, Jeanne L. "Reeling Pins," *The Spinning Wheel Sleuth* 12 (April 1996): 5-7.

Asplundh, Jeanne L. "The Search for the Elusive Girdle Wheel," *The Spinning Wheel Sleuth* 33 (July 2001): 2-3.

Bacheller, Sue. "The Sanfords of Newtown, Connecticut," *The Spinning Wheel Sleuth* 30 (October 2000): 2-4.

Bacheller, Sue. "Three Manufactured Swifts," *The Spinning Wheel Sleuth* 32 (April 2001): 9-11.

Bacheller, Sue and Florence Feldman-Wood. "S. Barnum and J. Sturdevant Double Flyer Wheels," *The Spinning Wheel Sleuth* 31 (January 2001): 5-7.

Baines, Patricia. *Spinning Wheels, Spinners and Spinning*. New York: Charles Scribner's Sons, 1977.

Baines, Patricia. *Linen Hand Spinning and Weaving*. London: B.T. Batesford, 1989.

Barker, Sister R. Mildred. *Holy Land: A History of the Alfred Shakers*. Sabbathday Lake, Maine: The Shaker Press, 1983. pages unnumbered

Barker, R. Mildred. "A History of 'Holy Land' – Alfred, Maine (Part I)," *The Shaker Quarterly*, Vol. III, No. 3 (Fall, 1963): 75-95.

Begnaud, Don. "The Acadian Spinning Wheel," *The Spinning Wheel Sleuth* 3 (September 1993): 4-6.

Betzner, Grant. "Update; Dutch Double-flyer Wheel," *The Spinning Wheel Sleuth* 7 (January 1995): 8-9.

Betzner, Grant. "The George Potts Spinning-Wheel Shop and the Moses Doolittle Wheel," *The Spinning Wheel Sleuth* 24 (April 1999): 8-9.

Bownas, Pat. "Picardy-Type Spinning Wheels," *The Spinning Wheel Sleuth* 27 (January 2000): 2-5.

Bownas, Pat. "How to Buy an Antique Spinning Wheel," *The Spinning Wheel Sleuth* 34 (October 2001): 9-12.

Bownas, Pat and David Bownas. "Two European Double-Flyer Wheels," *The Spinning Wheel Sleuth* 33 (July 2001): 7-10.

Brears, Peter C. D. "The York Spinning Wheel Makers," *Furniture History: The Journal of The Furniture History Society* XIV (1978): 19-24.

Buel, Cynthia B.B. *The Tale of the Spinning-Wheel*. Cambridge, USA: University Press, 1903.

Burnham, Harold B. and Dorothy K. Burnham. *'Keep Me Warm One Night': Early handweaving in eastern Canada*. Toronto: University of Toronto Press, 1972.

Buxton-Keenlyside, Judith. *Selected Canadian Spinning Wheels in Perspective: An Analytical Approach*. Ottawa: National Museums of Canada, 1980.

Channing, Marion L. *The Magic of Spinning*, 7th ed. Marion, MA, 1978.

Channing, Marion L. *The Textile Tools of Colonial Homes*. Marion, MA, 1969.

Conlin, Mary Lou. "The Lost Land of Busro," *The Shaker Quarterly* (Summer, 1963):44-60.

Coons, Martha with Katherine Koob. "Preindustrial Linen-making: The Process." In *All Sorts of Good Sufficent Cloth: Linen-Making in New England 1640-1860*. North Andover, Mass: Merrimack Valley Textile Museum, 1980.

Cummer, Joan. "The Double Flyer Spinning Wheel," *The Spinning Wheel Sleuth* 6 (September 1994): 2-4.

Cummer, Joan W. *A Book of Spinning Wheels*. Portsmouth, N.H.: Peter J. Randall, 1993.

Daniloff, Serge. "Some Rare Spinning Wheels." In *The Art of the Weaver*, edited by Anita Schorsch. New York: Universe Press, 1977. [First published in *The Magazine Antiques* (October 1929)]: 21-24.

Danner, Daniel. *The Account Book of Daniel Danner Beginning in the Year 1835*. Hershey Museum Archives, Hershey, Pennsylvania.

Davis, Nelda. "Great Wheel Tensioners," *The Spinning Wheel Sleuth* 13 (July 1996): 2-4.

Davis, Nelda. "Bishop Tabletop Patent Wheel," *The Spinning Wheel Sleuth* 26 (October 1999): 8-10.

Demming, D. "Account of the New York Accelerating-Head for Spinning Wool," *Archives of Useful Knowledge* Vol. II, No. 2 (October, 1811): 104-107.

Drepperd, Carl W. and Elizabeth Spangler. "The Spinning Wheel and the Art of Spinning," *The Spinning Wheel* (May 1957): 19-23.

Eliot, Doug. "Inside a Miner's Head," *The Spinning Wheel Sleuth* 29 (July 2000): 4-7.

Earle, Alice M. *Home Life in Colonial Days*. 1898. Reprint, Stockbridge, Mass.: The Berkshire Traveller Press, 1974.

Eaton, Allen H. *Handicrafts of the Southern Highlands*. New York: Russell Sage Foundation, 1937.

Esposito, Ralph J. "The Development of the Wool Spinning Wheel in the United States." Master's Thesis, State University of New York College, Oneonta at Cooperstown, 1970.

Evans, Craig F. "Elijah Skinner and Thomas Howland," *The Spinning Wheel Sleuth* 21 (July 1998): 2-5.

Evans, Nancy G. *American Windsor Chairs*. New York: Hudson Hills Press, 1996.

Evans, Nancy G. *American Windsor Furniture: Specialized Forms*. New York: Hudson Hills Press, 1997.

Feldman-Wood, Florence. "An Exhibit of Spinning Wheels: 'Handspinning in the Industrial Age: Patented Progress,' at the Museum of American Textile History," *Spin-Off, The Magazine for Handspinners* (Summer 1990): 66-73.

___. "Where are the Old Spinning Wheels?" *Spin-Off, The Magazine for Handspinners* (Summer 1990): 74-76.

___. "Swing Arms: From the Top," *The Spinning Wheel Sleuth* 1 (January 1993): 4-6.

___. "Swing Arms: From the Bottom," *The Spinning Wheel Sleuth* 2 (May 1993): 4-6.

___. "Two Wheelers: Turkish Wheels," *The Spinning Wheel Sleuth* 4 (January 1994): 4-5, 8.

___. "Update: Back on Tracks," *The Spinning Wheel Sleuth* 6 (September 1994): 7-9.

___. "Azel Wilder," *The Spinning Wheel Sleuth* 8 (April 1995): 7-8.

___. "Update – Solomon Dell's Lever Spinning Wheel," *The Spinning Wheel Sleuth* 11 (January 1996): 4-5.

___. "Vertical Two-Wheel Spinning Wheels," *The Spinning Wheel Sleuth* 15 (January 1997): 8-10.

___. "Amos Miner, Inventor of the Accelerating Wheel Head," *The Spinning Wheel Sleuth* 16 (April 1997): 2-3.

___. "Variations on a Theme: The Wheel-Head Collection of Sue Burns," *The Spinning Wheel Sleuth* 16 (April 1997): 5-7.

___. "A T.D. Brown Accordion-Arm Wheel," *The Spinning Wheel Sleuth* 23 (January 1999): 11-12.

___. "Guilford-Style Chair Wheels," *The Spinning Wheel Sleuth* 26 (October 1999): 2-4.

___. "Three Unusual Wheel Heads," *The Spinning Wheel Sleuth* 29 (July 2000): 8-9.

___. "Sanford Family Spinning Wheels," *The Spinning Wheel Sleuth* 30 (October 2000): 5-6.

___. "Double-flyer Spinning Wheels," *The Chronicle of the Early American Industries Association* Vol. 53, No. 4 (December 2000): 132-137, 171.

___. "Solomon Plant, Wheel Maker of Stratford, CT," *The Spinning Wheel Sleuth* 31 (January 2001): 2-4.

___. "Spinning-Wheel Maker List: List #3" *The Spinning Wheel Sleuth* (October 2001): 1-23.

___. "Bobbin Winder Basics," *The Spinning Wheel Sleuth* 36 (April 2002): 4-5.

___. "Three Unusual Bobbin Winders," *The Spinning Wheel Sleuth* 36 (April 2002): 6-7.

Franits, Wayne E. *Paragons of Virtue: Women and Domesticity in Seventeenth-Century Dutch Art*. Cambridge: Cambridge University Press, 1993.

Frey, Wilma. "Track Wheels in Iowa," *The Spinning Wheel Sleuth* 17 (July 1997): 9-11.

Frey, Wilma. "A Mennonite-Style Track Wheel," *The Spinning Wheel Sleuth* 23 (January 1999): 5-6.

Garvan, Beatrice B. and Charles F. Hummel. *The Pennsylvania Germans: A Celebration of Their Arts 1683-1850*. Philadelphia Museum of Art, 1982.

Goddard, Pamela. "Farnham Family Textile Tools," *The Spinning Wheel Sleuth* 10 (October 1995): 4-6.

Gordon, Beverly, *Shaker Textile Arts*. Hanover, New Hampshire: The University of New England Press, 1980.

Grant, Jerry V. and Douglas R. Allen. *Shaker Furniture Makers*. Hanover, N.H.: University Press of New England, 1989.

Gustafson, Laura. "Swiss Style Upright Wheel," *The Spinning Wheel Sleuth* 7 (January 1995): 3-4.

Gustafson, Laura. "Swiss Style Upright Wheels – Part II," *The Spinning Wheel Sleuth* 12 (April 1996): 8-9.

Gustafson, Laura. "Scandinavian Spinning Wheels," *The Spinning Wheel Sleuth* 14 (October 1996): 8-9.

Haynes, Elizabeth. "Mrs. Mifflin's Fringe Loom." In *The Art of the Weaver*, edited by Anita Schorsch. New York: Universe Press, 1977. [First published in *The Magazine Antiques*, October, 1945]: 28-29.

Henzie, Susie S. *After the Wheel, The Reel: Reels, Swifts, Winders*. Los Angeles, Ca.: 1998.

Hilts, Victor L. and Patricia A. Hilts. "Not for Pioneers Only: The Story of Wisconsin's Spinning Wheels," *Wisconsin Magazine of History* (Autumn 1982): 3-24.

Hilts, Victor L. and Patricia A. Hilts. "Why Patent Spinning Wheels: Some Additional Thoughts," *Spin-Off, The Magazine for Handspinners* (Spring 1996): 90-95.

Holcomb, Michael. "Signatures" *The Spinning Wheel Sleuth* 20 (April 1998): 10-11.

Holcomb, Michael. "A Lyman Wight Pendulum Wheel," *The Spinning Wheel Sleuth* 23 (January 1999): 4-5.

Holme, Randle. "Extracts from the Academy of Armory by Randle Holme of Chester, 1688," Excerpted by Alan Raistrick. *The Spinning Wheel Sleuth* 20 (April 1998): 2-5.

Holme, Randle. "Extracts from the Academy of Armory by Randle Holme of Chester, 1688 – Part II," Excerpted by Alan Raistrick. *The Spinning Wheel Sleuth* 21 (July 1998): 12-13.

Horner, John. *The Linen Trade of Europe* during *the Spinning-wheel Period*. Belfast: M'Caw, Stevenson and Orr, 1920.

Hubbs, Heidi. "A Dutch Double-Flyer Wheel," *The Spinning Wheel Sleuth* 33 (July 2001): 11-13.

Jeremy, David J. "British and American Yarn Count Systems: An Historical Analysis," *Business History Review* 15 (Autumn 1971): 336-368.

Keyser, Alan G., Larry M. Neff and Frederick S. Weiser, trans. and eds. *The Accounts of Two Pennsylvania German Furniture Makers: Abraham Overholt, Bucks County, 1790-1833 and Peter Ranck, Lebanon County, 1794-1817*. Breinigsville, Pennsylvania: The Pennsylvania German Society, 1978.

Kronenberg, Bud. *Spinning Wheel Building and Restoration*. New York: Van Nostrand Reinhold, 1981.

Leadbeater, Eliza. *Spinning and Spinning Wheels*. Shire Album 43, Shire Press, 1979.

Leggett, M.D., comp. *Subject-Matter Index of Patents for Inventions Issued by the U.S. Patent Office from 1790 to 1873, Inclusive*. Washington: Government Printing Office, 1874: 1393-1401.

Leinbach, William A. "Daniel Danner: The Man Behind the Wheel," *The Spinning Wheel Sleuth* 4 (January 1994): 6-7.

Lenderman-Kruse, Jane. "Two French Tabletop Spinning Wheels," *The Spinning Wheel Sleuth* 21 (July 1998): 6-8.

Lenderman-Kruse, Jane. "A Girdle Wheel by Fosters of Carlisle," *The Spinning Wheel Sleuth* 33 (July 2001): 4-6.

Lenderman-Kruse, Jane and Don Kruse "Tabletop Patented Spinning Wheels," *The Spinning Wheel Sleuth* 19 (January 1998): 2-5.

Mawhiney, Pamella. "A Wheel with a Bentwood Drive," *The Spinning Wheel Sleuth* 16 (April 1997): 8-9.

Mawhiney, Pamella. "Two 'Carriage' Spinning Wheels," *The Spinning Wheel Sleuth* 19 (January 1998): 11-12.

Mawhiney, Pamella. "A Distinctive Click Reel," *The Spinning Wheel Sleuth* 32 (April 2001): 8.

McMahon, James D. "Daniel Danner, Woodturner of Manheim, Lancaster County: An Early Nineteenth-Century Rural Craftsman in Central Pennsylvania." Master's Thesis, The Pennsylvania State University at Harrisburg, 1992.

Monkhouse, Christopher. "The Spinning Wheel as Artifact, Symbol, and Source of Design." In *Victorian Furniture: essays from the Victorian Society autumn symposium*, edited by Kenneth L. Ames. Philadelphia: The Society, 1982: 155-172.

Neher, Evelyn. *Inkle*. Guilford, Connecticut, 1974.

Nutting, Wallace. *Furniture Treasury*. New York: The Macmillian Co., 1928.

Packham, Jim. "A Small Scottish Wheel," *The Spinning Wheel Sleuth* 28 (April 2000): 5-6.

Palardy, Jean. *The early furniture of French Canada*. Translated by Eric McLean. Toronto: Macmillan of Canada, 1963.

Parslow, Virginia D. "Spinning Wheels," *Handweaver and Craftsman*, (Spring, 1956): 20-23.

Pate, Frank and Florence Feldman-Wood. "Patented Accelerating Spinning Wheel-Heads," *The Spinning Wheel Sleuth* 8 (April 1995): 5-7.

Pennington, David A. "John Green's Patented Spinning Wheel," *The Spinning Wheel Sleuth* 6 (September 1994): 4-5.

Pennington, David A. and Michael B. Taylor, "Wheelless Wheels,) *Ohio Antique Review* (July, 1988): 40-42.

Pennington, David A. and Michael B. Taylor. *A Pictorial Guide to American Spinning Wheels*. Sabbathday Lake, Maine: The Shaker Press, 1975.

Poppensiek, Neil. "Jesse Truesdell of Newark Valley, New York," *The Spinning Wheel Sleuth* 26 (October 1999): 6-7.

Raistrick, Alan R. "Some Thoughts on Reels," *The Spinning Wheel Sleuth* 28 (April 2000): 2-4.

Raistrick, Alan R. "Spinning Wheels from the Channel Islands," *The Spinning Wheel Sleuth* 28 (April 2000): 2-4.

Raistrick, Alan R. "A Manx 'Poppet' Wheel," *The Spinning Wheel Sleuth* 34 (October 2001): 2-3.

Ralph, Bill. "Canadian Tilt-Tension Production Wheels," *The Spinning Wheel Sleuth* 5 (May 1994): 6-7.

___. "Accelerating Spinning Heads," *The Spinning Wheel Sleuth* 8 (April 1995): 2-4.

___. "How to Identify Farnham Flax Wheels," *The Spinning Wheel Sleuth* 10 (October 1995): 2-3.

___. "Farnham Reels," *The Spinning Wheel Sleuth* 12 (April 1996): 4.

___. "Identifying Farnham Great Wheels," *The Spinning Wheel Sleuth* 13 (July 1996): 5-6.

___. "A Rare Farnham Accelerating Wheel," *The Spinning Wheel Sleuth* 15 (January 1997): 2-3.

___. "An Unusual Accelerating Head," *The Spinning Wheel Sleuth* 18 (January 1997): 9.

___. "An Unusual Dual Spindle Wheel," *The Spinning Wheel Sleuth* 19 (January 1998): 5-6.

Ramer, Barbara-Anne. "A Passion for Quebec Spinning Wheels," *The Spinning Wheel Sleuth* 34 (October 2001): 7-8.

Rettich, Hugo Edlen von. *Spinnrad-Typen*. Vienna: 1895.

Reynolds, William. "Some More Thoughts on the Vincent Spinning Wheel," *The Association of Ohio Long Rifle Collectors*, Vol. IV, Nancy G. No. 1 (February, 1982): 14-15.

Richman, Irwin. *Pennsylvania German Arts: more than hearts, parrots, and tulips*. Atglen, Pa: Schiffer Publishing, Ltd., 2001.

Rogers, Horatio. *S.D. Stevens and His Spinning Wheel Collection*. Salem, Mass. 1969 [reprinted from *Essex Institute Historical Collections* (January 1969)].

Schiffer, Margaret B. *Furniture and its makers of Chester County, Pennsylvania*. Philadelphia: University of Pennsylvania Press, 1966.

Schiffer, Margaret B. *Chester County, Pennsylvania Inventories 1684-1850*. Exton, Pa: Schiffer Publishing Ltd., 1974.

Schuck, Lois. "A Track Wheel in Missouri," *The Spinning Wheel Sleuth* 17 (July 1997): 8-9.

Seibert, Peter S. "Decorated chairs of the lower Susquehanna River valley," *The Magazine Antiques* (May 2001): 780-787.

Smith, John E. ed. *Our Country and Its People: A descriptive and biographical record of Madison County, New York*. Boston: The Boston History Company, 1899.

"The Spinning Wheel," *Ciba Review*, (December, 1939): 982-1016. (Reprinted booklet. Leeds: Rawdon Printing Co., 1977).

Taylor, Henry H. "Some Connecticut Yarn Reels," *The Magazine Antiques* (June 1930): 538-540.

Taylor, Janet Rognvaldson. "The Rognvaldson Spinning Wheel," *The Spinning Wheel Sleuth* 31 (January 2001): 11-13.

Taylor, Michael B. "Spinning Wheel Study Expands," *The Shaker Messenger*, (Spring 1986): 8-11, 23-27.

___. "Connecticut Chair Wheels," *The Spinning Wheel Sleuth* 5 (May 1994): 4-5.

___. "Deciphering Wear Marks," *The Spinning Wheel Sleuth* 18 (October 1997): 5-7.

___. "Marks on Shaker Spinning Wheels," *The Spinning Wheel Sleuth* 22 (October 1998): 2-4.

___. "Left-footed Wheels," *The Spinning Wheel Sleuth* 25 (July 1999): 4-5.

___. "Tensioning Devices on Shaker Great Wheels," *The Spinning Wheel Sleuth* 23 (January 1999): 2.

___. "The Pleasant Spinner," *The Spinning Wheel Sleuth* 27 (January 2000): 6-8.

___. "Two Unusual Reels – Marketing, Not Utility," *The Spinning Wheel Sleuth* 32 (April 2001): 6-7

___. "E.B. Sanford's Double-Flyer Wheel," *The Spinning Wheel Sleuth* 32 (April 2001): 13

___. "Humes, Danner, and Killian Flax Wheels," *The Spinning Wheel Sleuth* 35 (January 2002): 2-4.

Teal, Peter. "The Ranee Spinning Wheel," *The Spinning Wheel Sleuth* 36 (April 2002): 2-3.

Thompson, G.B., comp. *Spinning Wheels (The John Horner Collection)*. Ulster Museum, 1964.

Ulrich, Laurel T. *Good Wives: Image and Reality in the Lives of Women in Northern New England 1650-1750*. New York: Vintage Books, 1991.

Ulrich, Laurel Thatcher. *The Age of Homespun: Objects and Stories in the Creation of an American Myth*. New York: Alfred A. Knopf, 2001.

White, Frank. "Heads Were Spinning: The Significance of the Patent Accelerating Spinning Wheel Head," *Annual Proceedings of the Dublin Seminar*. Boston: 1999. pp. 64-81.

White, Frank. "Benjamin Pierce, Mid-19th-Century Wheel-Head Maker," *The Spinning Wheel Sleuth* 29 (July 2000): 2-4.

Wigginton, Eliot (ed.). *Foxfire 2*. Garden City: Anchor Press/Doubleday, 1973.

Wilson, Sadye T. and Doris F. Kennedy. *Of Coverlets: the legacies, the weavers*. Nashville, Tennessee: Tunstede, 1983.

Index

A Pictorial Guide to American Spinning Wheels, 4, 5, 122n
Academy of Armory, 88, 153
accelerating wheel, 6, 60, 61, 63, 65, 68, 69, 73, 78n, 90, 92, 95
"accordion" wheel, 104
American-Irish castle wheels, 53-59
"American Spinner", 110
American Windsor Chairs, 35
American Windsor Furniture, 38
American saxony, 9, 13, 30-39
Anderson, James, 135
Anderson, John, 135, 139
Anderson, William, 135
Antis, John, 150-151
arms, 10, 11
Asplundh, Jeanne, 153, 155
axle, 7, 11
Bailey, Benjamin, 49, 134
Baines, Patricia, 5, 11, 24, 43, 54, 56, 79, 86, 113n, 148, 151, 153, 157
Barnum, Silas, 79
bat's head, 16, 17
Bergeron, Louis Eloy, 157
Bishop, Lucius, 111, 112
Blackburn, Edna, 103
bobbin, 7, 9, 10, 11, 150, 159, 160, 162, 165
bobbin and flyer, 7, 10, 11, 12, 13, 14n, 30, 31, 40, 43, 125, 126, 128, 142, 145, 188
    bobbin lead, 11, 12, 38
    flyer lead, 11, 12, 31
    Picardy, 11, 12, 14n, 31
bobbin winders, 24, 25, 37, 168, 171n, 176
"boudoir" wheels, 148
box loom, see tape loom
Bracket, Isaac, 134
Briggs, Barnabas, 134, 139
Brown, T.D., 104
Bryce, John, 108-109
Burkhart, J. W., 107, 113n
"Burkheart (sic) patent wheel", 107,
Burley, John, 74, 75, 114, 157
Byrkit, Jesse, 101
Canada,
    New Brunswick, 102
    Nova Scotia, 38, 110, 111, 112
    Ontario, 22, 22, 101, 101, 102, 103, 105
    Quebec, 22, 37, 111
    Strathroy, Ontario, 101
Carr, Frances, 139
carriage wheels, 148-149
"cassel" wheels, 56, 114, 121, 124
castle wheel, 40, 53-59, 121, 122, 123, 128, 129n, 161, 166
Channing, Marion, 5, 171
chair wheels, 45, 48, 60, 66-70
Civil War, 98, 105, 113
Connecticut,
    Guilford, 66-68
    Newtown, 80, 83
    Stratford, 80
    Woodbury, 79
Connecticut chair wheels, see chair wheels
crank, 7
    external, 56-58, 70, 71
    internal, 48, 49, 50, 56, 57, 61, 62
Cummer, Joan, 5, 45, 70, 85, 101, 188, 189
Current, David, 101
Cushman, Thomas, 10, 134, 136, 139
Damon, Thomas, 165, 171n
Danner, Aaron, 121
Danner, Daniel, 10, 13, 54-56, 59, 114, 121
Danner, George, 123
Daniloff, Serge, 60
Dell, Solomon, 101
Demming, Davis, 92, 93, 97
Desmarets, Pierre, 111
DeWitt, Henry, 131
Diderot's Encyclopedia, 146
distaff, 5, 9, 11
    free-standing, 142, 173
    girdle, 173
distaff holder, 11
Doolittle, Moses, 105
double-flyer wheels, 35, 40, 43, 65, 66, 70, 73, 79-89, 114, 115, 120, 122n, 142
double treadle, single drive wheel, 45
double treadle, double wheel wheels, 40, 60-78, 96n, 115, 116, 120
Doughty, Joseph, 150-151
Draper, Nathaniel, 28, 134, 139
drive cord, 7
drive wheel, 6, 7, 8, 9, 11, 13, 15
drop spindles, 174
Edgerly, Josiah, 134, 139
England,
    Derbyshire, 153, 154
    Fulneck, XIII, 150, 151
    Leeds, 150, 151
    York, 150
Erb, John, 121
Ephrata Cloisters, 24
Esposito, Ralph, 134
European wheels, 6, 11, 13, 30, 31, 43, 88, 89, 140-157
Evans, Craig, 97
Evans, Nancy Goyne, 35
Fancher, W. 50, 50
Farnham, Charles, 117
Farnham, Frederick Augustus, 46, 117, 119
Farnham, Joel, 46, 96n, 98, 111, 115, 117-120
Farnham, Joel, Jr., 46, 117, 119
Feldman-Wood, Florence, 4, 80, 90
Fell, Frank, 36
Firmin, Thomas, 79
flax wheels, 9-11
    Scandinavian, 47
    Swedish/Norwegian, 46
    Swiss, 12
    Turkish, 96
Fletcher, Laodicea "Aunt Dicie", 113n
flyer, 9-11
footman, 9, 11
Foust, J. Wilson, 104
Fox (Fuchs), Jacob, 127, 128-129, 129n
Freeman, Nathaniel, 134, 135, 139
French-Canadian wheels, 6, 11, 31, 37, 50, 56, 78
Georgia,
    Rabun Gap, 24n
Germany,
    Kassel, 56
    Munich, 148
girdle wheels, 113n, 153-156
Goldsmith, Joshua, 66, 67
Goodrich, Edward, 134
Grant, Jerry, 139n, 171
great wheels, 7
Green, Irma, 50
Green, John, 98, 99
Greer, Nancy Osborne (aka "Granny Greer"), 21
Guthrie, Walter, 105
hackle, 171, 174
Hadd, Arnold S., 139n
Harding, Josiah, 137, 139
Hathorn, George, 110, 110, 112
Hawkins Brothers, 107
Henry, Samuel, 56
Hoit, Moses, 97
Holme, Randle, 88n, 113n, 153, 155, 156
Holmes James, 137, 139
Holmes John, 137, 139
Holmes Josiah, 137, 139
hooks, 10, 11
hoop rims, 7, 11, 30
Horner, John, 38n, 53, 56
Howland, Thomas, 98
Huguenots, 31, 56
Humes, Samuel, 32, 54, 96n, XV, 59, 122, 124, 128, 129n, 188
"hurdy" wheel, 111, 112
"inclined spinner", 98
Industrial Revolution, 111
"inkel lome", 169
inventories, 79
Irish castle wheels, 53-59, 121, 122, 123, 128, 129n, 161, 166
Jacobs, Jacob, 127
Jameson, John, 148
Johnson and Foust patent, 104
Johnson, James, 104
Johnson, Theodore, 139n
Kelia, W., 34
Killian, Jacob, 123, 129n
Killian, John Philip, 123, 129n
Kreger, Elizabeth, 113
Kunkel, Daniel, 20, 127
large wheels, 7
lazy Kate, 176
Leadbeater, Eliza, 122, 175
leathers, 9, 11
Leete, Edmund, 66
Leete, Eli, 66
Leggett, M.D., 96n
Leinbach, William (Bill), 4, 49, 53, 121, 129n
"Lever Spinning Wheel", 101
Libbey, Bennett, XII, 49
linen wheels, 7
long wheels, 7
Lyons and Lucas, 102. 103
"magazine rack" wheels, 146
maidens, 9, 11
Main, W.H., 108
Major, William J. 41, 129n
makers marks, 13, 14n; see Appendix for a list of makers and their marks.
    initials: (alphabetical order by first initial)
        "AOH", 124, 127, 129n
        "A.R.", 160
        "A.T.", 86
        "B.B.", 45, 70, 139
        "B.L.", 139
        "B.M.B.", 87
        "D.M.", 23, 131, 132, 137, 139
        "D.S.", 127
        "E.B.", 86
        "E.L.", 66, 67
        "F.W.", 33, 133, 134, 139
        "G.W.", 66, 68
        "H.M.", 95
        "H.S.", 127
        "I.H.", 129
        "J.A.", 139
        "J.H.", 137, 139
        "J.T.", 118
        "L. M.", 39
        "M.J.Y.", 76
        "M.R.", 78n
        "N.D.", 28, 134, 139
        "NF AL^D", 135, 139
        "N. G.", 44
        "N.M.", 137, 139
        "S.M.", 134, 139
        "S.P.", 80
        "SR AL", 23, 33, 134, 135, 136, 139
        "S.S", 80, 84, 88n 114, 122n
        "T.A.", 84, 85
        "T.C.", 23, 134, 136, 139
        "T.M.", 86
        "W+M", 41, 127
        "W.W.", 34
        "Z.W.", 33, 133, 139
    names: (alphabetical order by last name)
        "Alexander", 35
        "Barnum, S.", 79
        "Beers, Andrew", 85
        "Bird, D.", 81
        "Brown, L.", 46, 120
        "Bruce, J.", 109
        "Bryce, J.", 109
        "Burley, John", 74, 75
        "Calhoon, Ma.tt (sic)", 44

"Carter, T.", 36
"Creen, B." 127, 129
"Danner, Daniel", 10, 13, 54, 55, 56, 59,
"DOW", 161
"Edgerly, J.", 33, 134, 139
"Fancher, W.", 50
"Farnham, F.A.", 119, 120
"Farnham, J.", 46, 70, 72, 96n, 117, 118, 119, 120, 179
"Field, Russell", 81
"Fox, IA", 128, 129
"Fox, J.", 127, 128
"Frost, E.", 172, 173
"Heese, C.", 120
"Henry, S.", 50, 56
"Homsher, I.", 10, 125, 126
"Humes, S.", 32, 124
"Jacob, J.", 127
"Judson, L.", 81
"Kelia, W.", 34
"Killian, J.", 123, 129n
"Killian, P.", 123, 129n
"Klein, J.", 35
"Kunkel, D.", 20, 127
"L'Heureux, J.", 78
"Logan, W. H.", 43
"Mains, W.H." 108
"Merrill, Mel", 169
"Miles, J.", 66, 81, 98
"Paradis", 37
"Porter, C.", 28
"Reiner, D.", 125
"Reiner, S.", 125
"Rose, E.P.", 160
"Sanford, B.", 63, 80, 114, 115, 116
"Sanford, E.B.", 80, 82, 83, 84, 114
"Sanford, D.", 80, 114, 115, 122n
"Sanford, I", 80, 84, 88n, 114, 122n
"Sanford, J.", 80
"Schoonover", 120
"Sellers, I.", 125
"Stewart, W.M.", 56
"Sturdevant, J.", 79
"Thomson", 29
"Todd, A.", 125
"Truesdell, J.", 118
"Webster, A.", 61, 62, 63, 65, 84, 87
"Williams, E.S.", 28, 46, 70,
"Wood, M.", 59
Manuel du Tourneur, 157
Massachusetts,
  Boston, 32, 34, 95
  Essex, 69
  Lynn, 69
  New Bedford, 97
  Salem, 69
Matheny, Chelton, 36, 105
Mawhiney, Pamela, 52n
Mayville Furniture Company, 36
McMahon, James D. 123, 129n
Meacham, David, 131, 132, 139
Merrill, Nathan, 137, 139
Merrill, Stephan, 134, 139
Mennonite communities, 98
Michigan,
  Dundee, 36
  Grand Haven, 108-109
  Holland, 37, 38
  Traverse City, 36
  Zeeland, 36
Miner, Amos, 60, 78n, 90, 92, 96
Minor's head, 60, 61, 63, 96, 97-98
Missouri,
  Bethel, 100
  Cameron, 107
  Hannibal, 106
Moravians, 151
Morehouse, M.S., 102, 103
mother-of-all, 9, 11, 37
muckle wheels, 7
Museums and Historical Societies,
  American Textile History Museum, 5, 34, 62, 85, 97, 101 XII, 48,

Bayerisches Nationalmuseum, 148
Colonial Williamsburg, 151
Essex Institute, 69
Farmers Museum (N.Y.), 52n, 61
Henry Ford Museum, 74, 87, 98
Henry Whitfield State Museum (CT), 66
Hershey (PA) Museum, X, 42, 43, 44
Historic Lyme Village, 100
Home Textile Museum, 5, X, 42
Lancaster (PA) Historical Society, 56, 188
Museum of Appalachia, 113n
Pennsylvania Farm Museum, 55
Sandwich (N.H.) Historical Society, 97
Shaker Museums:
  Canterbury, NH, 131
  Hancock, MA, 131
  Old Chatham, NY, 131, 132
  Pleasant Hill, KY, 131
  Sabbathday Lake, ME, 131
The Heritage Society Museum of Lancaster County (PA), 123, 129n
Whaling Museum, 97
New Hampshire,
  Chesterfield, 93
  Keene, 87
  leg, 34
New York,
  Berlin, 49
  Brookfield, 90
  Cape Vincent, 102, 103
  Fergusonville, 63, 80, 114, 122n
  Greene County, 61, 115
  New Lebanon, 131
  Newark Valley, 118
  Old Chatham, 131
  Orwell, 120
  Owego, 18, 46, 51 119
  Tioga, 119
niddy noddy, 164
Norcross, L. 105
Nute, John Henry, 110, 111
Nutting, Wallace, 5, 60
Ohio,
  Bellevue, 100
  Hinckley, 100
  Holmes County, 18, 20
  Marietta, 36
  Milan, X, 41
  Washington County, 18
Oregon
  Aurora, 100
orifice, 11
Overholt, Abraham, XI, 45, 46, 124
Palardy, Jeanne, 56
parlor wheels, 14, 142-145, 150, 151
patent wheels, 40, 61, 90, 97-113
  borrowing of ideas, 105-112
  list of, 96n
  Patent Office fire, 83, 97, 105
Peabody, W.H., 97
pendulum wheel, 101-102, 113, 137
Pennsylvania,
  Berks County, 20, 27, 38, 41, 123, 127, 128, 130, 162, 180
  Bucks County, 26, 41, 123-125, 171n
  Carlisle, 56, 121
  Chester Springs, 40, 41
  Evansburg, 104
  Fayette City, 74,
  Lancaster, 32
  Lancaster County, 53, 54, 56, 59, 69, 123, 124, 125
  Landisville 53, 54, 59
  Liberty, 113
  Lycoming County, 74
  Manheim, 54, 114, 161
  Montgomery County, 26
  Philadelphia, 125
  Scranton, 101, 102
  Tioga, 118, 119
Picardy wheels, 11, 12, 14n, 31
Pixley, Anne, 5, 188

pitman, 7
Plant, Solomon, 80
Planta, John, 151-52, 157
Pleasant Spinner, 6, 90, 91, 92, 137
Potts, George, 105
Queen Victoria, 88, 89
quill winder, 24, 105, 106, 107, 168
Raistrick, Alan, 153
Ralph, William (Bill), 52n, 81
Read, Daniel, 90-92, 97, 98, 101
reeling, 159
reeling pin, 35, 38, 39, 54, 122, 126, 128, 160, 162
reels, 114, 115, 159, 161,171n,
  Connecticut, 161, 179, 180
  double dial, 162-163, 178
  drop arm, 178, 186
  European, 186
  five-armed, 121, 161
  four-armed, 115, 159, 161, 179, 180
  metal, 181
  New England, 159, 160, 180
  Pennsylvania, 121, 160, 161, 162, 178, 180
  six-armed, 115, 160, 161, 162, 178, 180
Rhode Island, 22, 85, 159
rimless wheels, 7, 24
Ring, Samuel, 33, 134, 135, 139
ripple, 175
Rogers, Charles, 97
rolags, 18, 26
Sanford (sic), E.B., 82, 83, 98,
Sanford, Beardsley, 63, 80, 114, 115, 120, 122n
Sanford, Elias Bristol, 80, 83, 84, 114, 120
Sanford, Isaac, 80, 83, 114, 122n
Sanford, Josiah, 80, 83, 114, 122n
Sanford, Samuel, 80, 114, 120
saxony, 13
  accelerating head, 95
  American, 9, 30
  Canadian, 37
  Connecticut, 35
  Dutch, 38
  Dutch-American, 37, 38
  double flyer, 88, 89
  eastern Pennsylvania, 35, 36, 38, 39, 40, 41, 123-129
  European, 30, 140-141
  German-American, 36
  "left-footed", 40, 41
  Low Dutch, 13, 30
  Low Irish, 13, 30, 38n
  metal, 111
  New England, 33
  New York, 33, 36
Schaffner, Gabriel, 121
Shakers, 13, 18, 23, 24n, 28, 29, 32, 33, 34, 78n, 84, 90, 91, 93, 96n, 131-139, 165, 181
  Alfred, Maine, 10, 23, 29, 33, 131, 133-136, 139
  Busro, Indiana, 137
  Canterbury, New Hampshire, 23, 33, 34, 90, 93, 131, 133-134, 136, 139
  Enfield, New Hampshire, 28, 133, 134, 139
  Hancock, Massachusetts, 133, 137, 165
  New Lebanon, New York, 23, 86n, 131, 132, 133, 137, 139
  Pleasant Hill, Kentucky, 18, 90, 131, 137
  Poland Hill, Maine, 23, 137, 138
  Sabbathday Lake, Maine, 23, 131, 133-135, 139
Shaw, Jacob, Jr., 100
Skinner, Elijah, 97, 98, 101, 113n
"slapper", 115, 120, 159, 179,
"small wheels", 7
spindle, 8
spindle holders, 17
spindle post, 8
spindle wheels, 7
  table model, 105
spinning wheel chair, 187

spinning wheel parts, 7, 8, 9
Spinning Wheel Sleuth, 5, 35, 189
Spinning Wheels, Spinners, and Spinning, 11, 79
spool knave, 176
Stevens, Samuel D., 5, 188
Stiles, George, 102
Stone, Robert G., 79
Sturdevant, John Jr., 79
Sturdevant, John Sanford, 79
swifts, 189
  basket, 168
  brass, 158
  scrimshaw, 165
  Shaker, 165
  squirrel cage, 166, 167, 168, 176, 177, 178, 187
  table top, 158
  "trillers", 166
  umbrella, 121, 137, 145, 165, 166, 177
table, 8, 9
table top wheels, 108, 109, 111, 112, 147, 149, 154, 155, 157, 158
tape looms, 169-171, 181-186
  box loom, 170, 182-186
  floor standing, 169, 182,
  two harness, 170, 181, 185
Teasdale, 153-156
Tennessee wool wheel 21
textile tool box, 172
Teter, David, 101
teeth, 11
tension devices, 7, 8, 9, 15-20
  flax wheels, 9, 30
  wool wheels, 8, 15-20,
    green-stick, 18
    rotating barrel, 18
    sliding table, 18, 19, 20
    under-the-table, 23
tow wheels, 51
treadle, 7, 9
Truesdell, Jesse, 120
Ulrich, Laurel Thatcher, 38n, 159
"vibrating frame", 100
"vibrating pendulum", 101
Vigeant, Seraphin, 111
Vincent, John, 18
Virginia
  Staunton, 107
  wool wheel, 20
Wait, Justin, 102, 103, 108,
Waiker, Walter, B., 104
walking wheels, 7
"weasel", XIV, 55
Webster, Alpheus, 45, 61-65, 73, 96n, 114-116, 120
Webster, James, 153
Webster, Robert, 153
West Virginia
  Beverly, 91
  Elkins, 91
wheel boys, 162, 163
wheel fingers, 162
wheel head, 7
wheel post, 7
wheel support, 7, 8
Wheeler, H.F., 98, 99, 100, 101
wheel-less wheels, 90-92
whorl, 11
Wight, Lyman, 101-103, 108, 113n, 137
Williams, Enoch Slosson, 28, 36, 46, 70, 117, 118, 120
Wilson and Fairbanks, 104, 106
winders, 159
Winkley, Francis, 133, 134, 139
Wisconsin
  Marietta, 108
  Mayville, 36
  Waitsville, 101
Wittse, A.S., 70, 118
wool cards, 171, 175
wool combs, 171, 175
wool wheels, 7, 8, 15-29
  children's, 24, 25, 26
  head, 7, 8, 17, 92-94
Wright, Zadock, 133, 139
yarn count system, 159, 171n